Environmental Leadership

Lincoln Filene Center
for
Citizenship and Public Affairs Series

Environmental Leadership
Edited by Stuart Langton

ηℓω

Environmental Leadership

A Sourcebook for Staff and Volunteer Leaders
of Environmental Organizations

Edited by
Stuart Langton
The Lincoln Filene Center for
Citizenship and Public Affairs
Tufts University

LexingtonBooks
D.C. Heath and Company
Lexington, Massachusetts
Toronto

Library of Congress Cataloging in Publication Data

Main entry under title:
    Environmental leadership.

    1. Environmental policy—Citizen participation. I. Langton, Stuart.
HC79.E5E5784   1984   363.7′057   80-7445
ISBN 0-669-03698-6

Published simultaneously in Canada

Printed in the United States of America on acid-free paper

International Standard Book Number: 0-669-03698-6

Library of Congress Catalog Card Number: 80-7445

To Todd, Heather, and Dean Langton,
lovers of wildlife, wilderness, and the bountiful sea

Contents

                 Stuart Langton   129

                 About the Contributors   139

                 About the Editor   141

Introduction

This book was written for people who are concerned enough about preserving the quality of our natural environment to do something about it. It is addressed to people who find themselves, whether by career, avocational choice, or unforeseen circumstance, in leadership or management positions in existing or emerging environmental groups. Among those who I hope might benefit from the ideas here are staff members of environmental organizations, active volunteers who assume office and other leadership responsibility in environmental groups, students of environmental affairs, and ordinary citizens who want to do something about an environmental problem in their community.

This is not a scholarly work, although a number of scholars have made contributions to it. It is primarily written by and for environmental activists. Its basic purpose is to provide general ideas and specific suggestions for people who engage in public education and advocacy to protect the quality of our natural environment.

This book is unique in two respects. First, these are all original essays that have been prepared by the authors specifically for this book. Second, almost all of the authors are successful and experienced environmental leaders. Lois Gibbs, for example, who wrote the chapter on community organizing, led the fight to protect homeowners at Love Canal. Allen Morgan, author of a chapter on fund raising, was for twenty years the director of the Massachusetts Audubon Society, the largest state environmental group in the nation. Brock Evans, author of the chapter on lobbying, has served as the lobbyist of the National Audubon Society and a state lobbyist for the Sierra Club. Nancy Anderson, author of the chapter on use of the media, is coordinator of the New England Environmental Leadership Network and Director of Environmental Affairs at the Lincoln Filene Center at Tufts University. Dr. Judy Rosener, who wrote about public hearings, has been a commissioner of the California Coastal Commission and wrote her doctoral dissertation on public hearings. Christy Foote-Smith, who coauthored the chapter on successful environmental-management practices, is executive director of the Massachusetts Association of Conservation Commissions. Mary Beth Bablitch, coauthor of the chapter on cable television, has been a staff aide to the Wisconsin legislature and a staff member of the Boston-based Citizens for Participation and Political Action. Douglas Amy, who wrote the chapter on environmental mediation, is not known as an activist, but his essay so impressed me with its balance and insight that I felt it should be included.

This book began more than twenty years ago, when Albert Bussewitz, an outstanding naturalist for the Massachusetts Audubon Society, stimulated my interest in natural history and conservation issues, as he had done with hundreds of others. Through his encouragement I became involved with the Massachusetts Audubon Society, serving as the director of the Moosehill Wildlife Society and later as director of the Ipswich River Wildlife Society.

Because I was inexperienced at the time, what I learned from those involvements had a profound effect on me. Since resources were very limited and public appreciation for environmental values (we called it conservation ethic in those days) was minimal, I quickly discovered the importance of volunteers, fund raising and management, public relations, and community organizing to a group devoted to environmental protection. I can recall wishing, however, there were a book to help me figure out how to do a better job in relation to these and the other leadership and management challenges.

In the late 1970s, I had three successive experiences that convinced me there was still much to be learned and said about leading and managing environmental organizations. As a volunteer officer of the Audubon Society of New Hampshire, I discovered how hard it was to be a dedicated and effective volunteer. When someone proposed building an oil refinery in my town, I rediscovered the importance and difficulty of fighting to protect the environment in one's own backyard. And when I received a grant to encourage environmental leaders throughout New England to network with one another, I found that there were hundreds of experienced and not so experienced volunteers and staff members of environmental groups who, like myself, were eager to find out more about managing and leading their organizations.

For six years, my colleague Nancy Anderson and I have had an opportunity to respond to these learning interests of environmentalists through the New England Environmental Leadership Network. While Nancy has done most of the work of organizing training institutes, conferences, newsletters, and meetings between government officials and environmental leaders, I have had the rare opportunity to learn through these activities—and from other leaders—some of the lessons that experience has taught me are necessary for effective leadership of environmental groups. This book is a modest attempt to record and share some of this learning. It is also an expression of gratitude to all those from whom I have learned, and an offering to those who might benefit from it in the future.

Acknowledgments

One of the most gratifying experiences in life is to have people help you to achieve what you could not have done alone. The preparation of this book has been such an experience for me.

To the authors who have prepared chapters for this book, I am deeply grateful. I asked these people to contribute to the book because they are so much more knowledgeable about the subjects they addressed than I am, and also because each of them is among the most experienced and successful people in the country in putting into action what they describe so well in words. Since they are all very busy people, I am all the more grateful for their contribution.

To Margaret Zusky and the editorial staff of Lexington Books, I can only marvel at their patience and tolerance while I completed this long-overdue manuscript. Their understanding was not only charitable, but inspired me to complete a project that, under other circumstances, I might have abandoned. I thank them for the gift of time.

To my colleagues at the Lincoln Filene Center for Citizenship and Public Affairs, I can only acknowledge that this work would never have been completed without them. Deborah Manning and Michael Woodbridge were extraordinarily helpful in their editorial assistance; and my associate and secretary, Virginia O'Neil, not only labored far beyond the call of duty but did so with a sense of equanimity and caring that I will always value.

1

The Future of the Environmental Movement

Stuart Langton

In 1963, former Secretary of the Interior Stewart Udall wrote *The Quiet Crisis,* which portrayed the environmental problems of the time and called for "a ground swell of concern . . . which could culminate in a third wave of the conservation movement."[1] In the years since, we have seen just such a "ground swell of concern" in changing public attitudes, the formation of countless environmental groups, and the creation of thousands of pieces of environmental legislation at all levels of government.

In fact, since 1963 we have witnessed the unprecedented success of the so-called third wave of the conservation movement. While only a few dozen national environmental groups existed in early 1960, there are over 350 groups today.[2] Although there was little effective federal legislation in 1963 to preserve the environment, by 1969 the National Environment Policy Act was passed, and by 1970 the Environmental Protection Agency was established. In 1965 only 17 percent of Americans identified the reduction of air and water pollution as a goal warranting government attention, but by 1970, 53 percent said government should be concerned with these issues.[3]

Many observers expected that the economic vicissitudes of the 1970s would derail or retard America's growing concern over environmental protection. However, that was not the case. Despite the threat of energy shortages, inflation, unemployment, and other forms of economic unrest, Americans continued to express concern about the quality of their environment. For example, a Harris poll conducted in 1978 indicated that 65 percent of the people disagreed with the proposal "to slow down environmental cleanup to ease the energy shortage," and 64 percent disagreed with the statement that we should "slow down the cleanup of air and water pollution to get the economy going again." Further, protecting the environment ranked higher (at 60 percent) as a national priority than creating hundreds of thousands of new jobs (44 percent) or decreasing dependence upon foreign oil (53 percent). A survey of 100,000 people in North Carolina during this same period further demonstrated the extent of public support for environmental protection in relation to economic development. In that survey, 85 percent of those polled thought it important to attract new industry to their state, but 89 percent felt it was also important to protect the environment.[4]

The election of Ronald Reagan in 1980 unnerved many environmental leaders who feared that the new president would weaken or undermine the

1

importance of environmental protection. The words and actions of Secretary of the Interior James Watt and Environmental Protection Agency Administrator Anne Gorsuch Burford, the stacking of federal agencies concerned with conservation and environmental matters with people who held antienvironmental views, and wholesale efforts by the Reagan administration to weaken or abandon environmental regulations confirmed the worst fears of environmentalists. However, the antienvironmental antics of the Reagan administration had several salutary spin-offs for the environmental movement. First, many environmental groups that had experienced some erosion in paid memberships because of the decline in the economy in the late 1970s suddenly experienced a swell in new memberships. Second, environmental groups, in the face of a common enemy, began to cooperate with one another as never before. Third, environmentalists discovered quickly how important it was to monitor existing environmental regulations and practices to protect their hard-fought victories of earlier years.

Beyond these unexpected benefits, the environmental movement also discovered that the American people had not abandoned their commitment to environmental protection despite the continued threats of the Reagan administration. A major public-opinion survey conducted for the Continental Group by Research and Forecasts in 1982 demonstrated clearly that, while there have been some changes in the public's perception of environmental problems, Americans remain strongly committed to environmental protection.[5] For example, 60 percent said that pollution is one of our nation's most important problems today; 84 percent felt that the balance of nature is easily upset by our activities; and 79 percent believed we should prevent the extinction of any animal. Further, the survey indicated that "Americans want both economic revitalization and (environmental) protection and, what's more, they believe they can have both."[6]

In addition, this study revealed two interesting trends in public attitudes. One was that the public perception of the intensity of the problem of air and water pollution had declined substantially between 1970 and 1980. Whereas 53 percent felt that this problem was one of three to which the federal government should devote attention in 1970, only 24 percent felt that way in 1980.

In probing this change further, the researchers found that the public "no longer sees protecting the environment as a 'crisis' issue; it is now widely perceived as a *management* issue."[7] The other trend was that business leaders were found to be much more supportive of environmental protection than previously believed. For example, 58 percent of the executives of large corporations and 45 percent of the executives of small corporations believed that the United States could achieve the goals of environmental protection and economic revitalization, and less than a third believed that we must relax environmental standards to achieve economic growth.

Such continued public support for environmental protection, despite the initiatives of the Reagan administration to weaken the environmental policies and conservation practices of the federal government, make clear that what was termed the quiet crisis of the environment is no longer a quiet matter. Environmental concern has become a significant aspect of American life. The third wave of the conservation movement, which Stewart Udall could only imagine twenty years ago, has become one of the dominant political realities of our time.

The Institutionalization of
Environmental Concern

In the relatively short period of twenty years, there has been a dramatic growth in institutions that reflect America's growing concern with environmental matters. For example, we have witnessed the passage of a federal Environmental Education Act and the creation of environmental-education programs and centers throughout the United States.

Environmental law, environmental education, environmental engineering, and environmental economics have become areas of professional specialization. Colleges and universities have developed courses, majors, and graduate programs concerned with the environment. New professional associations and journals have been created. The mass media devotes increasing attention to environmental issues in reporting and in special features and programs. Meanwhile, traditional conservation organizations such as the Audubon Society and the National Wildlife Federation have grown at an enormous rate, while many new national organizations such as the Sierra Club, Friends of the Earth, and the Cousteau Society have become prominent. At the local level, conservation commissions have been established in thousands of communities, and countless preservation groups and environmental action organizations have sprung up across the country.

This popular institutionalized success of the environmental movement has had a profound effect upon the United States. It has transformed how we think about our resources and growth. It has fostered awareness of issues not imagined two decades ago—such as toxic waste, acid rain, and carcinogenic pollution—and it has reoriented our economy inasmuch as environmental-protection projects now constitute one of its dominant growth sectors. It has led to the education of a cadre of new specialists, and it has created an entirely new political infrastructure that will remain for generations. As one commentator has noted: "Quite aside from these inescapable demands for reform, however, one can only marvel at the degree to which . . . concerns for water, air, noise, ugliness, and natural open spaces have entered the American consciousness, changed ways of living, and motivated the foundations of countless reform organizations—nearly all of this in a fifteen-year period.''[8] Indeed, this is a remarkable change within a relatively brief period of history.

The most distinctive feature of the third wave of the conservation movement has been the extent of its success. In the decade between early 1960s and early 1970s, this movement grew from a cause to an accepted set of institutionalized American values.

In the early 1960s, the term *environmental* was seldom used in a popular sense; but by the close of the decade it had become the popular slogan of a powerful movement. Americans became accustomed to hearing about environmentalists, the environmental movement, and environmentalism. By 1970, when the Environmental Protection Agency was created, the president of the United States referred to the 1970s as the environmental decade.

Beyond the Charismatic

Despite the success of the environmental movement, there are many who are uneasy about it because they feel that the character of the movement has changed. In the past few years I have spoken with many leaders of environmental organizations who fear that the movement has lost some of its momentum and vitality. As one leader remarked to me, "I think the environmental movement may be declining. I just don't see the new leaders. We may have run our course." Other leaders have expressed similar anxiety because of opposition and criticism by opponents of the environmental movement.

While the continued successes of the environmental movement hardly suggest decline, they do suggest change. However, this change does not appear to be engendered by failure as much as by success. What seems to be happening is that the environmental movement is being transformed from a relatively charismatic movement to a more institutionalized movement. It is a movement dominated less and less by charismatic leaders, protest, and dramatic causes as it becomes characterized by day-to-day efforts of education, research, and rule making. Furthermore, it is a movement less marked by inspirational leaders and more dominated by leaders who manage.

In these respects, the environmental movement in the 1960s and 1970s reaffirms Max Weber's analysis of how the charismatic is transformed into the routine dimensions of social and political life. As Weber noted in *The Theory of Social and Economic Organization*, if the charismatic "is not to remain a purely transitory phenomenon . . . it is necessary for the character of charismatic authority to become radically changed. Indeed, in its pure form, charismatic authority may be said to exist only in the process of originating. It cannot remain stable, but becomes either traditionalized or rationalized, or a combination of both."[9] Elsewhere, Weber obseves: "It is the fate of charisma, whenever it comes into the permanent institutions of community, to give way to powers of tradition or of rational socialization."[10] To Weber, such transformation of any popular movement was viewed as inevitable.

Today the environmental movement has become rationalized and routinized in relation to social institutions. It is now a movement that is essentially accepted and integrated into the mainstream of our cultural life. This is not to suggest that there is still no need for passion, drama, and charisma, but the environmental movement must confront the demands of rational socialization. In this regard it seems that a *fourth* wave of this movement is before us. The environmental movement has been successfully accepted; the new challenge is the management of that success.

New Competencies for Environmental Leadership

If environmental leaders and organizations are to be effective in their new institutionalized role, they need to develop skills of successful institutional leadership. This is as important for national and state environmental groups as it is for local ones. Therefore, the environmental movement must develop a dual consciousness to continue to be successful. There should be as much concern for the institutional capability of environmental organizations as there is for the substance of specific environmental issues. In the future, environmentalists need to be as concerned about the state of leadership of the environmental movement as about the state of the environment.

What areas of institutional capability are needed if environmental organizations and their leaders are to be effective? There are five areas of need in which environmental groups (or any public-interest groups) must strengthened competency in order to remain effective in the years ahead.

The Need for Professional Administration

Environmental organizations cannot afford the luxury of casual or sloppy management. The rigors of organizational survival in a complex institutional society demand efficient and enlightened administration. As David Cohen, former president of Common Cause, has warned:

> To succeed, a public-interest group must operate as a modern organization dealing with issues that matter to people, consulting with its members, focusing its constituencies and energies, respecting the professional role of the media by providing them with accurate and useful information, and

paying attention to administrative management. (Failure to pay attention to administrative management will snuff out all good intentions. The public-interest constituency will need its share of MBAs, just as it needs creative lawyers, imaginative researchers, skilled lobbyists, and inventive activists). . . . Professionalism must be a key style of operation for the public-interest constituency.[11]

While effective professional management is not a subject that is of particular concern to environmental leaders, it is important that they be attuned to it, as the quality of management will have a great deal to do with the advancement of environmental groups.

Professional administrative capabilities that particularly need to be fostered among environmental organizations include the ability to undertake moderate- and long-range planning, fund raising, financial planning and cost control, program evaluation, effective personnel practices, and management-information systems. The limited financial resources available to environmental groups require these skills for responsive stewardship; and the need to public trust calls for respected professional management.

Undoubtedly, some environmental leaders will fear that attention to professional administration may lead to bureaucratic excess in the environmental movement. Although this is an ever-present danger, it is not an inevitability. To avoid a cult of bureaucracy in the future, environmental organizations must be guided by values of humanistic management as well as efficiency. Therefore, environmental leaders who are recruited and trained should be as capable of working effectively with people as they are of handling administrative detail.

The Necessity of Collaboration

A dominant characteristic of modern organizational life is interaction among organizations. "Modern society," comments Amitai Etzioni, "has found it necessary to build more and more instruments to regulate this interaction to encourage increase not only in the effectiveness and satisfaction within each one but also of the relations among them."[12] This need for meaningful and cooperative relations among organizations is greater today than ever before, whether it be in the marketplace, in government, or in the voluntary sector. It is particularly critical for environmental organizations, since few possess sufficient resources to achieve their public education and sociopolitical goals.

Collaboration in the environmental movement must take place among environmental groups (intramovement collaboration) and with other groups that may share some common values (intermovement collaboration). Even though cursory review of the successes of the environmental movement gives evidence of the importance of both kinds of collaboration in the past, the present extent of this cooperation is not sufficient to sustain the success of the environmental movement in the future.

At an intermovement level, this lack of sufficient coalition building has become more evident in the past decade. Congressman Morris Udall noted several years ago that "part of the reason the environmental movement finds itself in trouble . . . is that we failed during the heady years of the 1960s to make friends and forge alliances with groups that might be largely with us now: blue-collar America, enlightened industry, the minorities who inhabit our rundown cities."[13]

In the years ahead, the environmental movement will have to form alliances continuously if it is to succeed. Accordingly, environmental leaders will have to develop appropriate attitudes and skills for coalition building. It will also be necessary to develop skills for permanent collaboration among environmental leaders and groups. Furthermore, these leaders will need effective skills in human relations and organizational development. They will have to design imaginative ways of sharing organizational resources. They will have to learn to build and sustain trust as they modify their organizational agendas to avoid duplication and unnecessary competition. They will have to learn how to ask for and receive help while learning how to give help. And above all, they will have to be able to create cooperative organizational systems in order to make collaboration a permanent, rather than an occasional, feature of their organizational functioning.

In order to foster collaboration among environmental groups, new types of organizational arrangements must be developed. For example, coalition organizations are needed at the local, state, and national levels to encourage joint planning among groups. Cooperative programs need to be established among groups in each state and region of the country to provide effective training opportunities for staff and volunteers. And joint efforts will be needed to identify and make available technical information and resource persons to environmental groups. Although creating these kinds of collaborative enterprises is difficult, it is absolutely essential in order to meet the emerging needs of the environmental movement.

The Importance of Political-Action Skills

Some years ago, John Gardner wrote: "Effective steps to save the environment will require a highly expert knowledge of government machinery, a knowledge of political infighting, a knowledge of how tough and enforceable legislation must be written. . . . These are subjects that well-intentioned Americans have stubbornly avoided, and by doing so they have all too often condemned themselves to failure in the battle to save the environment."[14] If future failures are to be minimized, environmental leaders, at all levels, need to develop these areas of political knowledge and skill. Three skill areas in particular will grow in importance in the years ahead: drafting legislation, lobbying, and participating in government-sponsored citizen-involvement programs.

Drafting Legislation. At the federal, state, and local levels environmental leaders should be informed about the process of drafting legislation, and how they can initiate, assist in, and influence this process. More specifically, environmental leaders need to be able to draft legislation and ordinances and work with appropriate legislative committees and their staffs. It is also important that they understand how to track the development of a piece of legislation in order to be able to support it.

Lobbying. While lobbying has always been an important political-action skill, the growth in the number of special-interest and public-interest lobbying efforts is changing the lobbying environment. Therefore, lobbying is becoming a very sophisticated art that requires considerable research and planning. Further, awareness of the nature and extent of lobbying permitted for environmental organizations that have tax exempt status under Section 501-C-3 of the Internal Revenue Code needs to be clearly understood.

Environmentalists must be concerned about the quality and quantity of their lobbying capability in light of the enormous growth in the lobbying efforts to business and industry. As an example of this growth, one need only consider the rapid increase in federally registered Political Action Committees (PACs). Between 1974 and 1980 the number of these groups grew from 608 to 2,551; whereas there were only 89 corporate groups in 1974, there were 1,204 in 1980.[15] This represents a 1300-percent growth rate in corporate PACs. Further, environmentalists should be concerned about the substantial increase in PAC contributions to political candidates. For example, in 1974 PACs groups gave a total of $12.5 million to congressional candidates, but in the 1980 congressional election the amount increased to $55.2 million, a 440-percent increase in six years. A 1976 U.S. Supreme Court decision in the case of *First National Bank of Boston* v. *Bellotti*, which allows business broad latitude in making political expenditures and lobbying, has further encouraged the growth of direct lobbying and political advertising by the corporate community. This means that environmentalists will have to increase the extent and the quality of their lobbying efforts to keep pace with the enormous increase in lobbying from private-sector interests.

Citizen Involvement. A major change in democratic practice in America in the past twenty years is the dramatic increase in citizen-participation efforts by government agencies.[16] As administrative agencies of government have increased their discretionary decision-making power, efforts to inform and involve citizens in agency decision making have become widespread. One study by the Community Services Administration catalogues hundreds of citizen-participation requirements and programs among federal agencies.[17] Another study has indicated over one hundred citizen-participation requirements in one state and estimated that as much as $50 million to $100 million may be spent annually in that state among state agencies for citizen-participation efforts.[18]

Although there is legitimate concern about the questionable quality of many citizen-involvement efforts, it is clear that citizen-involvement programs are here to stay and will provide important forums in which environmental groups should be involved. To participate effectively in these activities, environmental leaders must develop a number of citizen-involvement skills. Not among the least of these will be remaining informed of citizen-involvement opportunities at various levels of government, learning how to place capable representatives on advisory committees, preparing and delivering thoughtful testimony at public hearings, participating in shaping the rules for citizen participation in various government agencies, preparing enlightened written commentary, and monitoring citizen-participation programs to assure that the rules of the game are not violated and that citizen contributions are genuinely considered.

The Capacity for Scientific and
Technological Assessment

Nearly a decade ago, in his optimistic assessment of the potential relationship between man and technology, Victor Ferkis extolled the notion of a new model of what he termed technological man. "Technological man," he offered, "will be man at home with science and technology, for he will dominate them rather than be dominated by them."[19] Unfortunately, the day of technological man is still far off, because man seems hardly at home with or dominant master over science and technology.

It has been and will continue to be a principal mission of the environmental movement to assess the impact of science and technology on man and the biosphere. Because of the number and complexity of scientific issues and the rate of technological change, this is clearly an enormous challenge that environmental groups and organizations cannot meet by themselves. Further, it is unrealistic to expect any environmental group to be able to develop competency in more than a few areas of environmental concern. Therefore, it is important that environmental organizations develop imaginative organizational procedures for ensuring adequate scientific and technological assessment.

To ensure adequate and technological assessment, environmental organizations should develop capability in at least three areas. First, they will have to work closely with scientists and government officials to shape research and investigatory agendas concerning environmental issues. Second, environmental groups will need to collaborate in establishing a division of labor in regard to specific issues. To deal with issues in sufficient depth, different groups will have to specialize in one or several particular areas. Third, environmental leaders will have to know how to manage efforts of scientific and technological assessment in areas in which they have little or no technical training. This will require the knowledge and means of obtaining technical and scientific resources. Skill in attracting and assembling competent technicians and scientists, as well as dedicated citizens willing to study and learn (particularly on a voluntary basis), will also be needed. And it will require the ability to organize and help professionals and laypeople work together in the assessment of scientific and technical matters.

The Demand for a New Educational Orientation

H.G. Wells once commented that "human history becomes more and more a race between education and catastrophe." In the past fifteen years, the environmental movement has played a critical role in this race by increasing public awareness of catastrophic threats to our environment. In retrospect, it is clear that the educational impact of this movement has been monumental. It is clear that in the future this educational impact must continue.

However, environmentalists should anticipate new tensions and demands in their public-education role. In recent years, the environmental movement has been under increasing attack by critics as being essentially negative and obstructionist. In some instances these charges have been well-founded, and in others they have been mere propaganda. Nonetheless, these views are instructive to the future educational demands upon an institutionalized environmental movement. Basically, this suggests that the environmental movement must be as concerned with advocating realistic and positive solutions to environmental problems as in warning about environmental hazards. This means that environmentalists will have to gear educational efforts toward constructive discovery and advocacy as well as toward dedicated opposition. Without the former, the credibility of the latter will be minimized.

In this regard, environmental leaders will have to address issues of economic development from a balanced and well-informed perspective. This will require knowledge of economics and the ability to address critical issues with business and labor leaders in a constructive and realistic manner. A 1978 study of over eleven hundred environmental leaders in New England indicates a substantial interest in this area. For example, in rating forty-two issues, economics of natural resources protection (69 percent), economic development and the environment (58 percent), and jobs and the environment (52 percent) were rated as the first, fifth, and thirteenth issues of greatest priority among a majority of respondents.[20]

Expanding the educational focus of the environmental movement will not be an easy task, and many environmentalists may fall into the easy trap of adversary isolation. It is easier to be in opposition. It is easier to attract people to fight than to solve problems. It is easier to frighten than to enlighten. It is easier to arouse anger than dedication.

It is hoped that environmental leaders of the future will not take the easy way out, but will develop attitudes and skills to be constructive partners as well as tough opponents. It is hoped that they will be able to cooperate, create viable alternatives, and compromise when it is prudent. And, it is hoped, they will possess the human capacity to try to work with those with whom they most disagree.

The Environmental Movement in the Year 2000

What the environmental movement will be like in the year 2000 is uncertain. However, I suspect it will be highly diverse and decentralized, but much more strongly coordinated. Most likely it will be more institutionalized, although spontaneous protest movements will continue. It may well be that at least half of the issues with which environmentalists will be concerned in the year 2000 are things we hardly imagine today, and that the environmental movement will provide some of the most exciting career opportunities for the next generation. And I imagine that those who are committed to environmental values will constitute a major political force.

Above all, the environmental movement will probably be even more integrated into our social attitudes and institutions in the future than it is today. It is very likely that for the remainder of this century, issues of planning and political decision making will be as dominant as issues of technological development have been since World War II, and that anticipation and evaluation of environmental factors will be paramount. Further, I suspect that the tensions between economic development and environmental concern will be reduced and managed more effectively.

While this may bode well for the future of the environmental movement, it cannot happen unless new environmental leaders are attracted and unless these environmental leaders develop new competencies. If the fourth wave of the environmental movement is going to continue to swell in significance, it must be driven by the power of leaders capable of managing and leading environmental organizations with efficiency and effectiveness.

Notes

1. Stewart Udall, *The Quiet Crisis* (New York: Avon Books, 1963), p. 199.
2. *Conservation Directory* (Washington: The National Wildlife Federation, 1977).
3. Gallup Poll *Index* (June, 1970), p. 8.
4. *North Carolina Tomorrow: One State's Approach to Citizen Involvement in Planning for its Future* (Raleigh, N.C.: Governor's Office, State of North Carolina, 1978).
5. Kathy Bloomgarden, "Managing the Environment: The Public's View," *Public Opinion* (February/March, 1983).
6. Ibid., p. 47.
7. Ibid., p. 48.
8. Sydney B. Ahlstrom, "National Trauma and Changing Religious Values," *Daedalus*, Journal of the American Academy of Arts and Sciences (Winter, 1978), p. 22.
9. Max Weber, *The Theory of Social and Economic Organization* (New York: The Free Press, 1966), p. 364.
10. Max Weber, *From Max Weber: Essays in Sociology*, ed. and trans. by H.H. Gerth and C. Wright Mills (New York: Oxford University Press, 1967), p. 253.
11. David Cohen, "The Public Interest Movement and Citizen Participation," in Stuart Langton, ed., *Citizen Participation in America* (Lexington, Mass.: Lexington Books, D.C. Heath and Co., 1978), p. 60.
12. Amitai Etzioni, *Modern Organizations* (Englewood Cliffs, N.J.: Prentice-Hall, Inc., 1964), p. 113.
13. *Congressional Record* (July 16, 1975).
14. John Gardner, "Foreword," in Nelson Rockefeller, *Our Environment Can Be Saved* (New York: Doubleday and Co., 1970).
15. David Cohen, "Who Says PACs Have to Own the Congress," *Citizen Participation* (September/October, 1982).
16. Stuart Langton, ed., *Citizen Participation in America* (Lexington, Mass.: Lexington Books, D.C. Heath and Company, 1978).
17. *Citizen Participation* (Washington, D.C.: Community Services Administration, 1978).
18. Stuart Langton and Associates, *A Survey of Citizen Participation Requirements and Activities Among Major State Agencies of the Commonealth of Pennsylvania* (Harrisburg, Pennsylvania: Stuart Langton and Associates, 1976).
19. Victor C. Ferkis, *Technological Man: The Myth and the Reality* (New York: George Braziller, Inc., 1969), p. 246.
20. Stuart Langton, *Learning Interests and Needs Among Environmental Leaders in New England* (Medford, Mass.: Lincoln Filene Center for Citizenship and Public Affairs, Tufts University, 1978).

2 Dilemmas of Leadership and Management in Environmental Organizations

Stuart Langton

In chapter 1, and throughout this book, the term *leadership* is used in a general sense to refer to a number of activities that environmental organizations need to undertake to be successful. Although the term *management* is also used often, the two terms are not interchangeable. There are some subtle and some not-so-subtle differences between leadership and management. This distinction, I believe, is very important for environmental organizations to understand because it clarifies some of their most common needs and difficult problems.

The Difference between Leadership and Management

The history of management and leadership theory is rich with definitions and disagreements. Nonetheless, an examination of the etymological roots and variations of the terms *leadership* and *management* suggests a relatively clear difference in their meanings. The Anglo-Saxon *laedan* for example, from which the term *leadership* is derived, means to travel, to go, to move in some way, to set a direction. The term *manage*, particularly in French (from which it has passed into current use), means something quite the opposite. The French *ménage*, for example, means housekeeping. A related word, *manège*, an equestrian term, means to bring under control, as in reining in and controlling a horse.

These distinctions suggest that leadership and management represent two very different functions of organizational life. Leadership relates to directing or moving the organization to achieve some purpose or to serve some value. In this sense, leadership points out the need for change and attracts people to support it. "Leadership," observes James MacGregor Burns, "is exercised when persons with certain motives or purposes mobilize . . . resources so as to arouse, engage, and satisfy the needs of followers."[1] Management, on the other hand, is concerned with maintaining order over the resources and practices of the organization. As Henri

Foyol, a French engineer, pointed out in his pioneering work on the subject in 1916,[2] management includes such administrative functions as planning, organizing, supervising, and controlling. Whereas management looks primarily inward in its efforts to control, leadership looks outward in its efforts to bring about change.

Although these functions are quite different, they must both be performed if an organization is to be successful. As Alvin Gouldner points out, organizations require some people to perform leadership functions and others to perform management functions. In referring to these different roles as agitators and bureaucrats, rather than leaders and managers, Gouldner observes: "In the modern association, the needs which both agitator and bureaucrat serve may be neglected only at the risk of probable impairment to the association. Both, or functional equivalents of them, seem necessary to the modern democratic group."[3]

Gouldner's point is particularly relevant to environmental organizations that have a membership base and are involved in advocacy. Such groups have a strong need for leadership, particularly in promoting values and change regarding environmental protection and preservation. At the same time, their management needs are great because they must administer membership records, renewal notices, publications, finances, and the like. If they don't take care of their management needs they will not have resources to promote their values; but, if they don't promote values, they will not attract resources.

**Changing Needs for Leadership
and Management**

One of the fascinating characteristics of organizations is that at different times in their history they need to devote more attention to leadership than management, and vice versa. This was an important observation of Max Weber, who believed that movements and organizations move through successive cycles, beginning with a period in which they are inspired by a charismatic-type leader who founds or reforms it. This stage is followed by one in which the practices of the organizations are no longer dependent on the personality of the leader, but are based instead on rational management principles; and, finally, these principles are converted to bureaucratic rules and procedures that provide the institution with greater order and control.

In many ways, environmental organizations reflect patterns of growth and change similar to that described by Weber. At some point, environmental groups need a strong leader to build or reform the organization. It is the ability of such a person to articulate values, create new programs, and attract a following of supporters that brings the group success. The paradox of this for persons who have performed this function for environmental groups is that their very success creates a new series of problems for them and the organization. The reason is that once followers are attracted or increased, additional resources are created and new activities and services are made available; then, administrative procedures are needed to manage and maintain these new features of the organization. Some leaders are very good at building an organization to this point, but then they are either not interested or incapable of managing the success for which they were largely responsible. In fact, some of the greatest leaders of the environmental movement have either quit or been fired as a result of their success as leaders and their failure as managers.

A common pattern in environmental organizations facing this dilemma is to hire a good manager to put the so-called house in order. However, while a person may perform this function well, his or her success may also create problems in that while they were managing the organization efficiently, they may not have been able to give it enough direction. Consequently, the nonprofit world in general—and the environmental movement in particular—has seen many good managers kicked upstairs or out the door because they were not taking their organizations anywhere.

Constructive Transitions in Leadership and Management

While some organizations handle these changing demands for leadership and management well, all too often these situations create conflict and trauma for the members and staff of the organization. In many cases the board becomes split down the middle, staff is demoralized, and supporters are lost. Can these situations be avoided? Is it possible for environmental organizations to address changes in leadership and management needs constructively? I believe that they can if board members and staff are cognizant of these inevitable transitions in an organization's development and if they create mechanisms to deal with them.

As a consultant to a variety of organizations, I have observed five things that can be done to cope constructively with these shifting needs. First and most important, an organization should assess, from time to time, how well it performs its leadership and management functions. A board-and-staff retreat is often a good time to do this. A consultant may be very helpful in assisting with this process or in conducting a study of the organization to clarify its changing needs. These activities should point out whether the organization needs to devote greater attention to either its leadership needs or its management needs before it gets into serious trouble.

Second, the expectations that a board of directors has regarding the leadership and management priorities of their executive director should be made explicit. If these priorities change over time, they should be discussed, and an executive should be given time to shift gears if he or she is capable of doing so. When hiring a chief executive of an environmental organization, the board should recognize that there are very few people who possess outstanding leadership *and* management skills. Those who do are in short supply and great demand. Further, many executives who are best at leading become bored with management, and many who are best at managing are uneasy with the risk taking and public exposure that is associated with leadership. Therefore, a board needs to consider the strengths and shortcomings of an executive in relation to the needs of the organization at a particular time.

Third, hiring an executive for a specified period of time is one way that some groups get the best person to deal with either pressing leadership or pressing management needs. Many talented executives, in fact, like to take on a specific challenge for a designated period of time and then move on. In many fields, there are executives well known for their ability to turn a situation around or launch an organization because of their unusual leadership abilities; and, there are those who have a reputation for housekeeping because of their superior management skills.

Fourth, many executives can grow in their jobs and develop new competencies. Sometimes a good leader can become a better manager and a good manager can become a better leader. This often happens in small and moderate-sized environmental organizations, since most of the people who are hired to direct them have limited experience. I have found two practices to be very helpful in this kind of situation. One is that the executives meet annually with a consultant, the board president, and/or the executive committee to discuss their needs and plans for professional development. (As a result of this, I have seen very successful and experienced board members provide helpful consultation and advice to these less experienced executives.) The other practice is to join or create a group of executives for the purpose of mutual training.

Fifth, many organizations simply understand and accept the unique strengths and limitations of their chief executives, but they appoint other persons to perform functions that the executive does not do well or chooses not to do. For example, in some organizations the executive director serves primarily as a goal setter, spokesperson, and fund raiser, and an associate director manages internal affairs. In other organizations the director serves primarily as a chief administrator while many of the leadership functions are carried out by associates who are responsible for such things as public relations, lobbying, and financial development.

Tensions between Leadership and Management

Because leadership and management functions differ, it is not uncommon for tensions to arise over these two needs in organizations. These tensions can frequently be observed among staff, within boards, and in individuals. For example, staff members whose responsibilities are primarily administrative often become upset with what they perceive as the poor management practices or insensitivity of program leaders. Program leaders become upset with administrative staff for what they perceive as nit-picking and failure to appreciate the importance of the greater goals of the organization. Administrative staff often perceive program leaders as acting like chickens with their heads cut off while program leaders perceive administrative staff as being uptight.

Similar tensions arise on boards. For example, board members who are preoccupied with the finances and budget of the organization become uneasy with those who want to talk about new programs; and those who are primarily interested in programs and issues are impatient with the cautiousness of the business types. Some individual staff members, such as chief executive officers, also experience these tensions within themselves. For example, a director of an environmental organization may feel torn between wanting to testify at a public hearing and needing to prepare a budget report for the board. Whichever is chosen, the person usually feels guilty or uneasy about not having done the other.

Although these kinds of tensions are often irritating, the fact of life is that they are inevitable. In every organization there is a constant battle between leadership needs and management needs—a continuous struggle between the bottom line and the front line. The important thing is not to let these tensions get out of hand, not to let people become antagonistic or bitter about them, and not to allow one of these needs to overshadow the other.

What can environmental organizations do to create a healthy balance and tension between their leadership and management needs? While there are some practices and procedures that can be helpful, I believe that the key to integrating those two needs is attitudinal. Above all, staff members, whatever their functions, need to appreciate both the managerial and leadership demands on their organization and show respect for those who perform functions that differ from theirs. The contributions and problems of everyone from the janitor to the president need to be appreciated; and the person who is most responsible for setting this tone within the organization is its director. By example and through supervision, a director should not allow condescension or bickering, two very common human tendencies, to go unchecked. Whenever conflicts arise—as they inevitably do—between people who are responsible for different functions, the director should encourage attitudes of openness, caring, humor, and compromise.

Beyond this, there are several concrete things that can be done to encourage a healthy balance between leadership and management functions. These include:

1. *Hold an annual retreat for staff.* This event can be used to inform staff about each other's functions, discuss and resolve conflicts that have built up over the year, and increase appreciation of the leadership and management needs of the organization. Such a process can help staff to understand and accept the inevitable tensions they experience, and it provides an opportunity to agree on ground rules and procedures for reducing conflicts.

2. *Conduct regular staff meetings.* On a regular basis, staff should come together to inform each other of plans and discuss existing or potential problems. However, a good staff meeting should be more than a reporting session. It should be a time when those responsible for developing and directing programs alert administrative staff of potential or expanded programs. Administrative staff should provide feedback about their capacity or limitations in regard to potential—and existing—programs and problems. As a result of discussing these matters, the entire staff should attempt to reach agreement on how to reduce existing problems and how to proceed with or alter new programs.

3. *Establish policies and procedures.* Tension is often reduced and conflicts avoided when people know the rules. The problem with many organizations is that they have no clear rules about how to proceed. One way to overcome this problem is to create a policies-and-procedures manual that establishes rules for leadership and management functions. It is particularly helpful to have a team of staff members who represent several functions write a draft of the manual, and for all staff to help review and revise it before it is adopted. In this way, all staff become aware of everyone else's functions, the rules everyone has to honor, and the procedures for interfacing their functions.

4. *Require administrative staff to review program proposals.* A very common practice in environmental groups and other nonprofit organizations is for program staff or fund-raising specialists to write program proposals without having them reviewed by administrative staff. If these proposals are later funded, it is often found that the proposal may contain unrealistic plans, inadequate budgets, or unreasonable requirements for administrative staff. To avoid conflicts that arise from this practice, an organization should require that all proposals be reviewed by the treasurer or comptroller, secretarial supervisors, administrative

assistants, publications director, or any other management specialists who will be responsible for any aspect of the proposed program.

5. *Require program staff to review administrative procedures.* As organizations grow larger, new rules and procedures are frequently developed by specialists hired to perform specific managerial functions. These rules may address such varied matters as personnel policies, travel, parking, and record keeping. While bureaucratic procedures may seem necessary to one manager, they may seem shortsighted, unnecessary, or downright stupid to other staff. In addition, they may undermine the capacity for people to conduct programs or perform other functions. To avoid such excesses of bureaucracy, an organization can require that no procedures be adopted without a staff review.

While the principles and practices suggested above may reduce the tensions between people whc perform leadership and management functions in organizations, they will not do away with them. What these approaches can do is to maintain a more healthy balance between competing organizational needs by encouraging people to participate in relation to each. The process is a continuous and creative one that requires an attitude of constructive cooperation.

Notes

1. James MacGregor Burns, *Leadership* (New York: Harper and Row, 1978), p. 18.

2. Henri Foyol, *General and Industrial Management* (London: Pitman Co., 1949).

3. Alvin Gouldner, ed., *Studies in Leadership* (New York: Harper and Bros., 1950), p. 64.

3 One Hundred Ideas for Successfully Managing and Leading Environmental Groups

Stuart Langton and
Christy Foote-Smith

This chapter reports the results of our asking two dozen successful leaders of environmental organizations to share some of their most important learnings in leading and managing their groups.

Although environmental organizations vary considerably in size, structure, purpose, and activities, there are many management and leadership needs they have in common, such as planning, fund raising, and lobbying. In conducting interviews with our sample of leaders, we identified fifteen of these issues and asked each person to describe the best approaches to meeting each need and to point out problem areas.

Our compilation of some successful stratagems is a reflection of many years of collective wisdom. As many of the leaders advised, it is important that environmental activists not waste their time reworking already-established basics; they should instead learn from each other's successes and mistakes. Herewith are some of those learnings.

The Importance of Planning

The leaders who participated in this survey stressed the importance of planning as the first step to effective management. Critical to this process is the need to establish a clear mission statement, a list of long-range goals, and a list of well-defined annual operating objectives. These procedures are very important, notes Kelly McClintock, executive director of the Environmental Lobby of Massachusetts, so that "organizations will not get sidetracked."

This chapter reflects the results of a research survey undertaken by the authors in 1981. The survey covered 24 well-known executives of environmental organizations with outstanding reputations among their peers. The sample includes leaders of national, regional, and local organizations from various sections of the country. However, the sample is weighted heavily toward leaders of Washington-based national organizations and organizations in the Northeast, since these areas have the highest concentration of environmental organizations.

The survey was based on in-depth telephone interviews conducted by Christy Foote-Smith. An interview questionnarie was jointly designed by the authors. The interview findings were assembled by Christy Foote-Smith and the final narrative was prepared by Stuart Langton.

A mission statement, it was frequently pointed out, is very helpful because it clarifies the basic purpose of an organization. This is important, since many groups court trouble by trying to be all things to all people. A mission statement forces a group to declare what it is as well as what it is not. In preparing a mission statement, several leaders offered two key suggestions. One is to be certain that the statement is consistent with the needs and concerns of the group's members. The other is to make the statement neither too broad nor too narrow.

In addition to a mission statement, an effective organization needs to be clear about its priorities. The best way to do this is to develop a list of long-range goals. Organizational leaders can use these goals to select activities annually that best serve the goals. But a long-range plan serves another practical function. As Sue Ellen Panitch, president of the Connecticut River Watershed Council, points out, "a long-range plan is the best selling tool of the organization to new members and funders." Tom Arnold, director of the New England Rivers Center, adds, "a long-range plan can be translated into a brochure. Foundations want this material."

What is the time frame for a long-range plan? Leaders of larger environmental groups suggest three to five years is an acceptable norm; however, as Douglas Foy, executive director of the Conservation Law Foundation of New England, warns, "Five-year plans are hard to stick to since things change." For groups with no membership base, it was suggested that two years is a good time frame for which to plan, since these groups must be very entrepreneurial and adapt quickly to changes in their funding environment.

Annual planning, most of the leaders recommend, consists of selecting a series of objectives that a group is to achieve each year to serve its goals. (The difference between a goal and objective is discussed in Stuart Langton's chapter, How to Write a Proposal.) In addition to objectives, a group's annual plan should include a list of activities, a schedule, and a budget. A budget is a critical part of this process, since it determines how the organization will invest its limited resources. Who should prepare the annual budget? Leaders were divided on the question. Some prefer to have staff do it, since they know more about the organization than most; while others felt a board committee, consisting of people with financial-management experience, was effective. Nonetheless, there was one common denominator in both views: the preparation of the annual plan requires a great deal of communication, cooperation, and trust between board and staff.

In developing annual plans, three recommendations were made. The first is to involve members in planning. One method for doing this, suggests Brock Evans, vice-president of the National Audubon Society, is to "poll the membership. This builds support and trust. It also provides an opportunity to educate them.'' An annual meeting or retreat is another method used by many groups. Maggie Foster, president of the Alabama Conservancy, reports that their board ''gets together in March for a two-day planning session to identify issues, set priorities, and work on the internal issues of the organization.'' If staff do most of the planning, Judy Kiriazis and Larry Kamer (director and assistant director of the Lake Michigan Federation) suggest that staff, ''check board-meeting minutes and identify the consistent issues. Have they been adopted as programs? Have you dealt with all the issues set forth the previous year?''

A second recommendation is to be realistic in planning. ''Set goals you can achieve. Objectives can provide guidance if they are realistic. Budgets help if they are realistic,'' advises Rafe Pomerance, president of Friends of the Earth. Roland Clement, president of the Connecticut Audubon Society, adds, ''You must know how much will be available in resources to get the job done.'' ''The budget,'' warns Douglas Foy, ''should not be a 'wish list,' but a realistic plan.''

A third recommendation is to avoid rigidity and be flexible. ''Goals are important, but rigid adherence would be a mistake,'' suggests Kelly McClintock. Effective management, according to many leaders, includes striking a balance between being clear about where you are going and being open to change. ''Plan in advance as much as possible,'' points out Tom Kimball, executive vice-president of the National Wildlife Federation, ''but leave plenty of elbow room. Things can happen fast. You must work for your objectives, but must also have the resources to put out fires.'' Brock Evans adds, ''Environmental groups cannot control events, and must be prepared to react. You can't always set the agenda.'' Acknowledging this reality, Nancy W. Anderson, director of environmental affairs at the Lincoln Filene Center for Citizenship and Public Affairs, recommends, ''Don't be too set in your methods. You must be able to reassess and change priorities.''

Administration

A point upon which almost all of our leaders agreed was that environmental organizations need to pay more attention to good business practices, and that few environmental leaders have practical training in this area. ''Most of our leaders are not trained professional managers. We came up through the ranks fighting battles, and then we became heads of organizations. We don't know very much about personnel evaluation, and hiring and firing, and meeting a budget, and all the other trappings of management,'' comments Brock Evans.

Despite this, the leaders offered a number of recommendations to cope with this need within environmental groups. One suggestion is to encourage staff who are not experienced in management to attend seminars and take courses. However, many leaders noted that it is often difficult to train someone to be a good manager who has little prior experience or inclination to learn. "You can spend a lot of your limited resources trying to make an administrator or manager out of someone who is not," notes Harry Miller, president of Trout Unlimited. "The best thing for the organization, and for that person, to do is to change their position or terminate them and bring someone else in who has the skills you desire."

Two clear trends are that environmental groups are placing greater emphasis on management skills in hiring new staff, and that groups increasingly use consultants to assist with such administrative functions as long-range planning, accounting, and data processing. "In hiring a consultant," suggests Tom Kimball, "make sure you get someone who knows something about your field, and can take a fresh look at what you're doing and make suggestions as to how you can do it more professionally."

Some groups, it was reported, use board members to assist with management. Several leaders suggested using members and volunteers, particularly if a group is small. Among the areas in which members can make important volunteer contributions are bookkeeping, legal matters, accounting, membership-recordkeeping, research, conference management, and preparation of publications.

A common problem that emerged is that environmental groups are not changing mangement practices as they grow. "Many groups start small with very relaxed practices," observes Janet Welsh Brown, executive director of the Environmental Defense Fund. "As they grow, their responsibilities get larger, but management practices don't grow." This is a serious problem for many environmental groups. As Douglas Foy points out, "Growth can be painful, and restructuring to accommodate growth generally meets with resistance."

In discussing management issues, many leaders referred back to the importance of planning. "The most important method of cost control," claims L. Gregory Low, executive vice-president of The Nature Conservancy, "is a good long-range plan." Tom Deans, executive director of the Appalachian Mountain Club, adds, "A good long-range plan provides a good tool for saying no to marginal new programs. You learn to cut the expendable things, not the things that are critical to the long-range success of your organization." Building on this principle, a number of leaders suggested developing procedures to monitor a group's activities in relation to its long-range and annual plans. "Not meeting your objectives," says Tom Kimball, "should always trigger a review."

Several additional suggestions for improving the quality of administration were mentioned a number of times by our group of leaders. These include: (1) Preparing job descriptions for staff and volunteers. This is increasingly important as a group becomes larger. (2) Communication. This is essential. Staff and volunteers should be aware of organizational plans, policies, and problems. Also, staff should meet regularly (weekly or bimonthly), and supervisors should encourage or give constructive criticism to staff. "Lack of feedback," suggests Tom Deans, "is a breakdown in effective management." (3) Keeping an eye on the larger issues. A group should not get so bogged down in its internal management concern that it forgets its larger purpose. (4) Not overmanaging. "Share decision making. Rely on delegation. Spread responsibility around," advises Rafe Pomerance. Nearly all the leaders stressed the importance of working as a team and developing skills of team management as a way of being more productive and of keeping morale high.

Financial Management

In discussing finances, many leaders stressed the importance of realistic budgets, thorough records, ongoing methods of reviewing finances, and controlling costs. In general, there was a keen awareness of the impact of inflation on environmental groups. Consequently, four related points were frequently made. The first point was that every organization must live within its means. "Keep a surplus, no matter what your income," advises Harry Miller. "Unless you can keep yourself with a positive fund balance, you're not attractive to outside donors."

A second point was to monitor income and expenditures constantly. "What you need more than anything else," claims Gerard Bertrand, president of the Massachusetts Audubon Society, "is good information on what you're spending your money on." In this regard, it was pointed out repeatedly that an organization's treasurer or bookkeeper plays a very critical function in informing the organization of its financial status. However, people in these positions can only be as effective as the record-keeping systems a group develops.

A third point was to recognize the importance of a budget and to use it as a management tool. "Reducing or controlling costs is done primarily through budgeting," suggests Rafe Pomerance. "That's how people know what they've got to spend, and they know that they have to stay within those limits." Several leaders reinforced this ethic by suggesting that it is important to inform and involve all staff in budget preparation and review.

The fourth point, which several leaders mentioned, was that environmental organizations have much to learn from the business community. Douglas Foy points out, "We really have to view our operation as a business. Keeping costs down is no different for us than it is for a small corporation. It requires close management, attention to detail, and careful book-keeping, accounting, and auditing procedures, which I don't think many environmental organizations have paid attention to."

Several specific suggestions for controlling costs were identified. Among these: (1) Train all new staff in cost-saving techniques. (2) Use volunteers whenever possible. (3) Don't become top-heavy; spend less on staff and administration and more on programs. (4) Always attempt to have equipment donated rather than buying it. (5) Seek free printing from a printer or company that can receive free publicity from their donation. (6) Share resources with other groups whenever possible. (7) Use bulk-mail rates for large mailings. (8) Reduce phone costs by installing WATS lines or other similar bargain long-distance systems. (9) Fly tourist class or on special charter rates when possible. (10) Insulate your building to reduce energy costs.

One final point that a number of leaders mentioned concerned staff salaries, the largest expense item for most environmental groups. Several leaders acknowledged the fact that people will work for environmental organizations for less compensation than usual because of their personal values. It was also noted that there has been a surplus of such people for a decade. Nonetheless, it was suggested that it is unwise to expect excessive cost savings because of these factors. Gerard Bertrand observes, "One of the ways in which you cannot save very much is in salaries of technically qualified professionals. It would be better to have fewer employees who are real pros than having a larger number of semiqualified people." Michelle Perreault, vice-president of the Northeast Region of the Sierra Club, ob-serves further, "Make sure, where staff are concerned, that you are not paying them so little that they end up leaving. Then you have to hire again and you're just spinning your wheels."

Fund Raising

Raising money for environmental groups is a task that draws spirited and dispirited responses from leaders. While almost all leaders recognize its im-portance to a group's survival, few of them relish it. "It's mostly a lot of hard work," "It doesn't happen by magic," "There are no silver bullets," and "I wish I didn't have to do it" are some typical responses.

Many among our sample pointed out how important it is for environ-mental leaders to improve their skills at fund raising because of the growth in competition for the philanthropic dollar. "Money is just not available that was available a decade ago," observes David Zentner, board chairman of the Izaak Walton League of America.

Our leaders offered a veritable potpourri of fund-raising methods that included such traditional things as charging membership dues; receiving grants from government, businesses, and foundations; individual solicitation; bequests; annual fund drives; direct-mail campaigns; and income from publications sales and program admission fees. Then there were a number of innovative activities suggested, including art auctions, rock concerts, field trips, lectures, lotteries, and raffles. Several leaders commented that direct mail had become increasingly popular in recent years, but warned that there were many pitfalls in using this method because of its high initial cost and because so many environmental groups are conducting direct-mail campaigns. "It is like the tragedy of the commons," one person observed, "We are overgrazing."

Three general guidelines were mentioned frequently. One is to build diversified income sources and not put all your hopes into one method. (One group follows a rule of not relying on any one source for more than 5 percent of its income.) However, most groups find that membership dues are the most stable and diversified scurce (see Allen Morgan's chapter, on Fund Raising for Environmental Groups, for further discussion). While most leaders warned about the dangers of relying on a few large donors, one observed that cultivating large donors provides the greatest return for the least amount of time and money expended.

A second guideline is to recognize that providing good programs is the best way to attract income. "Fulfill the dictates of your charter," advises Gerard Bertrand, "and a lot of money will come in by itself." As an organization builds a successful track record over time, it will increase its potential for attracting income. Nonetheless, Sue Ellen Panitch recommends that a group continuously perfect its pitch to potential donors. In so doing, she suggests that packaging a group's appeal as a rallying point over a critical issue is extremely effective.

Third, the personal face-to-face approach in fund raising was deemed crucial. "People don't give money to causes, but to people," asserts one leader. Accordingly, it was suggested that personal visits to meet the staff of foundations, corporations, and government agencies, which might provide grants, was as important as writing a proposal. While some leaders felt it was crucial for the chief executive of an environmental group to visit potential funding sources, Michelle Perreault adds, "The people who are giving the money should come in contact with the 'excitors'—those who have conceived and executed projects."

In reflecting on their fund-raising experiences, several leaders added one word of caution. Increasing income does not happen overnight, and there is usually an eighteen-month lag time in receiving funds from new sources. So, groups should not wait until they are in trouble to increase their fund-raising efforts. Anticipatory planning is a necessity.

Proposal Writing

One of the most interesting findings of our study was that many of our sample had little or no experience at proposal writing although all had experience in fund raising. Those who had experience suggested the following points.

The most frequent suggestion was that groups learn a great deal about a funding source before submitting a proposal. Obtain written reports and talk to people about a foundation or corporation before you submit a proposal. Make personal contacts, if possible, with staff and board members of a foundation or company. Robert H. Gardiner, Jr., executive director of the Natural Resources Council of Maine, advises, "You should know before you write the proposal that you are going to get the money."

To those who have little experience at proposal writing, it was suggested that they obtain copies of successfully funded proposals, find some publications on the subject, and attend a seminar. In dealing with an R.F.P. (Request for Proposal) from a government agency, several people pointed out the importance of reading the requirements very carefully and completing all forms exactly as instructed.

Who should write proposals for an organization? The responses we received varied. However, there was a general consensus that those closest to a proposed program should write the proposal. "The team that will implement the project should write the proposal," recommends John H. Adams, executive director of the Natural Resources Defense Council. Harry Miller adds, "This is left to the professionals on the staff or board, or a hired consultant. We cannot afford to be amateurish."

Among other tips offered were: keep proposals concise, reasonable, specific, and well defined; have a realistic budget; do not propose to do something that has been done or that someone else is doing; and don't invest too much time in preparing the proposal.

Obtaining and Renewing Memberships

Not all environmental organizations have paid members. However, those that do find them to be vital to their success. "The key to success," says Tom Kimball, "is to keep your constituency, because they are a sound source of assistance, financially and politically."

How does a group attract new dues-paying members? The most common response among those we interviewed to this question was that the issues a group deals with are of the greatest interest to prospective members; the issues must therefore continue to appeal. "My experience," claims

Brock Evans, "is that issues bring people. If you are out in front on some kind of an issue, you are bound to attract people to your banner."

Among other inducements to attract members is offering good programs, services, or products. For example, the Appalachian Mountain Club has built a strong following for over a century on this basis. "We have never used any direct mail in order to get members," comments Tom Deans, "We use our reputation as the means of attracting membership through our facilities and publications." As for products, many leaders believe that publications, especially newsletters or magazines, are an important incentive to members.

Leaders of larger environmental organizations made it clear that direct mail has become the single most important way to obtain new members. A number of leaders pointed out that direct mail is expensive and complicated, and recommend that dependable and successful consultants be used. "If you send out a million pieces," comments Brock Evans, "and get a 1-percent return, that is ten thousand new members. That's very effective. It's down to a science." Leaders of smaller groups suggested several other methods of attracting new members, including holding special events, sponsoring conferences, and setting up displays at fairs and shopping centers.

Almost all the leaders stressed how important it was to entice members to renew their memberships, and offered suggestions very similar to those provided by Allen Morgan in the following chapter. However, two additional suggestions were offered. One was to personalize correspondence to all members whenever possible. Use terms such as *you* and *I*, and use a typewriter that allows you to use the member's name even in form letters. The other suggestion was to conduct occasional surveys to determine why members did or did not renew.

Board Development

A board of directors can make or break an organization, depending on what it does. Although boards vary in terms of how much influence they exert on an organization, according to our leaders, one principle is clear: While board members may perform a great many management and program functions as volunteers in small groups, in moderate-sized or larger organizations a board should deal with policy and keep its hands out of management. "In larger, more permanent organizations, the board should be a policy body. Administration should be separate," advises Janet Brown.

What constitutes a good board? Many of our leaders state that members should possess at least two of the following: wealth, work, or wisdom. "Try to get a balance of talents," recommends L. Gregory Low. Among the types of people our leaders recommended are: lawyers, technical people, scientists, business executives, grass-roots activists, accountants, academicians, well-known people, government officials, and proven fund-raisers. "Don't put token luminaries on the board," warns Douglas Foy. But, Tom Kimball adds, "People of reputation give credibility, and this puts pressure on people you want to influence." "Look for people without huge egos," recommends Brock Evans.

How can good board members be identified and recruited? Our leaders suggest that personal contacts are important. "Find a few good people, and you will attract others," says L. Gregory Low. Harry Miller suggests that "you need a vigorous nominating committee to look for people who wish to be involved." Staff and board members can assist in this process by keeping a list of people who they feel have talent and interest.

Developing an effective board is no easy task. "Getting work out of a board is a lot of work," notes Kelly McClintock. Several ideas were suggested for eliciting the most from a board. "You must be clear about what you want from them," says Janet Brown. Tom Deans recommends, "Have job descriptions." Several leaders believed that a strong president who knows how to delegate work and demand results from members is critical. Training sessions for board members are also extremely important.

Selecting and Managing Staff

Because most environmental organizations are relatively small and staff are often expected to perform many functions, the selection of a new staff person is a substantial investment. Almost all of our leaders suggested that a selection process be thorough. "Go slowly," advises Kelly McClintock. "Choosing the wrong person can be a big problem." In searching for new staff, a number of suggestions were offered: set up a strong search committee, advertise in professional journals, interview carefully and completely, scrutinize resumes, and follow up on references.

Many leaders emphasized that there is no substitute for selecting people of high quality, even if it costs more. L. Gregory Low recommends, "There should be a stress on excellence from top down: high intelligence, commitment to the purpose of the organization, a good conservation ethic, and an interest in producing results. You must get people who are team players." In a related point, Roland Clement advises, "Don't put too much trust in paper qualifications. You must be comfortable working with the person. Experts are no good if you don't get along."

One point many of our leaders made is that there are serious problems with the job market in environmental organizations. These include low pay, high turnover, a glut of qualified candidates, and few opportunities for advancement. "The chains of command are short. The chances for advancement are low. This can lead to a lot of frustration and burnout," comments Gerard Bertrand.

These factors, combined with the strong emotional investment that staff usually have in addressing environmental issues, call for supportive and creative management. "It's important to keep morale high since salaries are low," suggest Judy Kiriazis and Larry Kamer. Team building is extremely critical, many of our leaders said. "Try to have a staff that is close, with a warmth of feeling and common sense of purpose," advises Douglas Foy. "Camaraderie is key."

Utilizing Volunteers

Volunteers can be of enormous assistance to environmental organizations. In fact, many smaller groups carry out most of their work through volunteers. Although most of the leaders we talked with had served as volunteers themselves and had managed volunteers, almost all noted that there were substantial problems in utilizing volunteers.

"Supervision of volunteers is a critical factor in using them effectively," suggests John H. Adams. But, as Janet Brown observes, "Professionals don't supervise volunteers well." In addition to problems of supervision, volunteers are frequently not trained well, and this can result in a volunteer misrepresenting or reflecting poorly on the organization. These problems, it was pointed out, are caused primarily by staff who are not trained in volunteer management and will not take the time necessary to supervise volunteers. Despite these problems, the leaders of several groups in our survey indicated that they found volunteers to be very helpful in fund raising, legal research, lobbying, maintenance work, and publications and programs assistance.

Obtaining volunteers does not seem to be a problem for most environmental groups. "We have no need to recruit," reports Gerard Bertrand. "People just come to us." Douglas Foy adds, "We get more applicants than we need." However, many of our leaders suggested that it takes effort to find well-qualified volunteers. "Always keep eyes and ears open to good potential volunteers. Cultivate them, and keep a resource list," recommends Tom Deans. Many leaders report that students and retired persons are usually good volunteers; and many groups find student interns to be very helpful.

Planning, training, supervision, and recognition were identified as four key necessities in effective volunteer management. Before volunteers are recruited, staff should identify activities that can be performed by volunteers. To do this, several leaders suggested that staff break down each program or management function into smaller segments and determine which ones volunteers can carry out. Once this is done, qualifications can be identified, and a volunteer job description can be prepared. Several people recommended that volunteers identify their abilities and interests on a form, and this could be used to match volunteers with the jobs that need to be done.

The motivation and needs of volunteers should be clearly understood. "Give them meaningful work," advises Brock Evans; "They need to feel satisfied," comments Tom Deans; "Give them leadership roles," recommends Rafe Pomerance. Although volunteers need to feel that their work is important, Nancy Anderson warns, "don't be afraid to ask them to help with the nitty-gritty, like addressing envelopes and licking stamps." Above all, many of our leaders said, volunteers need to feel important and appreciated. Among ways of doing this, they suggested that staff express respect for volunteers, and that their efforts be recognized in newsletters and annual meetings, at special luncheons and award ceremonies, and through letters of thanks, certificates, and gifts.

Effective Publications

Publications are the major image pieces of environmental organizations. "Publications," notes Kelly McClintock, "are the only way, by and large, that members and the general public have of knowing who you are and what you do." Recognizing this fact, our leaders emphasized that a group should prepare publications that reflect high quality in content and style.

There are many things that make a publication effective, but one of the first things is to be very clear about its purpose. While this sounds like a simple maxim, several leaders pointed out that many environmental groups do not do this. Accordingly, it was suggested that a group determine the extent to which a publication is a vehicle for educating people about issues, entertaining them, encouraging them to act, or informing them about the organization. In relation to this, it is important to know what readers of a publication want, and how they judge the publication. "Make sure that publications are pleasing to your membership," suggests Tom Kimball. "Use readership surveys. Ask what people like and don't like about your publications."

Because environmental groups vary in size, our survey indicated that publications are managed in many different ways. Large organizations tend to employ specialists. Others hire consultants and part-time technical people, such as graphic artists. Smaller groups often appoint one staff member to spend a portion of his or her time on publications, rely on volunteers, or get an advertising agency to help them on a *pro bono* basis.

Some groups have a volunteer editorial-review board or a publications committee of the board. But, whatever procedures are used, it is suggested that senior management always review and approve what is being published.

What are some of the elements of a good publication? Our leaders offered many suggestions: use a crisp style, be imaginative, be consistent, and use attractive graphics, illustrations, and photographs. "Be precise, be relevant, and give complete and up-to-date information," suggests Nancy Anderson. "Publications should challenge people's minds and contain new documentation," advises Michelle Perreault. "Don't try to squeeze too much material into too small a space. Don't use words that are too long or concepts that are too obscure," says Robert Gardiner. When it comes to newsletters, several people suggested that frequency is more important than length, although this may increase cost. Cost, it was noted, is an important issue affecting quality. To control costs, four specific suggestions were offered. First, establish and maintain deadlines. Otherwise you will pay more in printing and postage than you need to. Second, always use bulk-mail rates. Third, do not feel the need to use the most expensive paper in multicolored printing. Fourth, shop around for printers, and send out work on competitive bid.

Selecting and Monitoring Issues

One of the ongoing problems of environmental leaders is remaining informed about issues with which their organizations are concerned. Because there are so many important issues, several leaders urge groups to select a few issues they can deal with well rather than becoming too diverse. It was suggested that groups be aware of the issues that others are addressing and not overlap unless it is a part of a strategic and cooperative plan. Otherwise, as Douglas Foy suggests, "You must be able to dig out problems no one's paying attention to."

To remain informed about environmental issues, five suggestions were offered. First, a group's membership can help to identify and monitor issues, use it. This can be done by sending out an issue questionnaire to members, or by establishing one or several committees. Second, select a reasonable number of publications and devise a plan to have some reviewed regularly by staff members or volunteers. Costs can be reduced if groups exchange magazines and newsletters. Third, develop a personal-contact network so that knowledgeable people can keep you informed. "I get better information from personal contacts than from reading," says Kelly McClintock. Fourth, attend meetings and conferences with other organizations. Fifth, keep in touch with the opposition by reading their materials or meeting with their leaders.

Organizing for Action

Almost all environmental organizations engage in programs of public eduation and advocacy. It is far easier, noted many leaders, to generate public interest and mobilize people to oppose something than it is to support a positive course of action. Also, the public and policymakers hear about so many issues that only well-organized action campaigns can compete successfully for their attention. For these reasons, almost all of our leaders commented that thorough research and strategic planning are imperative.

It was recommended that one person be responsible for conducting issue research. This person can use and coordinate the efforts of other staff, volunteers, or consultants. In addition to obtaining the complete facts about an issue, including relevant scientific and technical data, it was suggested that a political profile be undertaken to identify the positions and plans of proponents and opponents. "You must know your subject. The opposition will have money and will be educated and credible," point out Judy Kiriazis and Larry Kamer. To counter this, it is important to know the arguments of the opposition and to develop more compelling counterarguments. These arguments should consider economic as well as environmental factors. Gerard Bertrand advises groups to propose constructive alternatives: "One of the biggest faults of the environmental movement has been the failure to come up with an alternative that is reasonable or helps solve the problem." Since public policies are frequently a matter of compromise, several leaders suggest that groups identify fall-back positions as well as preferred alternatives.

Once a group has done its homework and is clear about its position, it must develop a well-conceived strategy. "Most good strategies are not single strategies," suggests Janet Brown. Therefore, a group should carefully select a variety of means for informing, involving, and influencing people. The group should also plan how to coordinate and orchestrate its activities at the outset. "The secret of political power is in the hands of people working together, not individuals working alone," observes Brock Evans.

Our survey of leaders offered several helpful tips for organizing: "Cast the issue in as personal or as local a vision as possible," suggests Kelly McClintock. "Getting media attention is the key," advises Robert Gardiner. "It's like training a mule. You have to hit him over the head with a two-by-four." Several leaders pointed out that educating and mobilizing the public begins with education of the organization's membership. "We do this through meetings, passage of resolutions, a newsletter, or an urgent letter outlining what members can do," reports Robert Herbsp, executive secretary, Trout Unlimited. "Have a sizable number of your members personally write letters to politicians illustrating that the individual knows something about the subject," advises Tom Kimball. Some leaders mentioned the importance of timing. "The most effective time to organize opposition is when a problem is first discovered or a proposal is made," states Nancy Anderson. "It is twice as hard to be successful once a financial investment is made or when a plan seems locked in concrete."

A final piece of advice offered by most of our leaders was to seek out allies and form a broad base of support. Coalitions of environmental groups were strongly recommended. Also, groups should look outside the environmental community, and should particularly seek out business interests that might be hurt by an environmental problem. "Seek the broadest possible base," suggests Tom Arnold, "because very few projects are defeated just because environmentalists oppose it."

Lobbying

Because most environmental organizations have received nonprofit tax-exempt status from the Internal Revenue Service, there are legal restrictions as to the amount of lobbying in which they can engage. Consequently, some groups are fearful of lobbying. This is unfortunate, observes Kelly McClintock, since "there is an immense amount that they can do and that would be very effective in furthering their own purposes without remotely infringing upon their tax status." Several other leaders agreed with this view and suggested that every group should understand federal and state laws regarding lobbying and, having done this, to make the most of their right to lobby.

Five points were mentioned frequently in discussing how to lobby effectively. The first is that the best lobbying is done at the grass roots since legislators are accountable to their constituents and want to be reelected by them. "Calls and letters from constituents is what really counts in turning out votes," observes Janet Brown.

Second, it is important to know and understand whom to lobby. Members of key committees and their staffs need to be identified, since these are the people who most influence legislation. Once these people are identified, it is helpful to understand their backgrounds, points of view, and constituent base before approaching them.

Third, whenever possible, make personal contacts with those you lobby. This includes legislators as well as their staffs. While letters and phone calls convey a point, a case can be made more effectively face to face. Be sure to thank the legislator who is supportive. Those who are particularly helpful might be given awards by an environmental group.

Fourth, when meeting with legislators, it is essential to present your case well. "You must know what you're talking about. Public officials can spot people who are just repeating information," advises Roland Clement. In particular, several leaders suggested that you counter the arguments of the opposition during meetings with legislators.

Fifth, always remain objective and credible. "Rule number one is never get mad. The friendly approach always works best," advises Robert Gardiner. "Maintain credibility. Never lead a public official astray or oversell your case," recommends Tom Deans. "If you get a reputation for not being trustworthy, you're dead in your tracks." In presenting an objective position, a number of leaders suggested addressing moral and economic questions as well as technical environmental matters.

Testifying at Public Hearings

Several leaders suggested that although public officials are the main audience to whom public testimony is directed, environmental groups should also use public hearings to influence the media and the general public. "Get all the mileage you can out of your testimony," urges Michelle Perreault. This can be done by writing public testimony that will also be of interest to the press, to group members, and the general public. In addition, a group can send out a press release and a fact sheet at the time of the public hearing.

In preparing testimony, many of the principles of lobbying are applicable. Among the specific suggestions offered were: be completely accurate, avoid being emotional, don't be windy or repetitive, keep it relatively short— and remember people can remember only three to five facts, prepare an additional background paper of greater detail than your talk, make clear why your group is interested in the issue being addressed, and be certain you represent the views of your organization accurately.

In regard to who should testify, several ideas were expressed. One was that no single group should testify on behalf of a coalition because of the difficulty in getting a consensus from all coalition members on all the fine points. Further, testimony from all members of a coalition has a stronger political impact. Another idea is to have a respected expert in the field testify on behalf of an organization. A final suggestion was to have a series of questions given to each hearing officer or legislator before the hearing so that they will see you are knowledgeable and responsive to these key questions.

Obtaining Technical Information

Whether it is for purposes of public education or advocacy, environmental groups need to obtain adequate scientific information about their issues. "It is essential to have technical backup," asserts Tom Arnold. "For any environmental issue, the position must be supported by evidence." Failure to obtain such data can lead to inadequate or incorrect positions and to loss of credibility. "Many groups don't get the technical information they need before they make decisions on policy," observes Gerard Bertrand. "This causes a lot of backlash and leads to a lack of understanding of issues, or to focusing on the wrong issues."

A few larger environmental organizations have scientists on their staffs who can undertake research or are knowledgeable about current research in selected fields. Some organizations can afford to hire scientists on a consulting basis. But most groups either rely on a volunteer advisory committee of scientific and technical experts or utilize staff, volunteers, or student interns with the background and ability to obtain necessary background information from existing sources.

Among the technical sources that several leaders reported to be of most help were libraries, the National Academy of Science, congressional studies, and reports of government agencies. College reference librarians and professors are often very willing to identify helpful sources of information. In addition, industry and consulting organizations may have conducted research that they will share. To save time and energy in identifying good source material, it was suggested that a number of other environmental organizations be contacted, and that computer-literature searches be undertaken.

Using the Media

Almost all our leaders acknowledged that use of the media was one of the best tools for public education and political advocacy, but that few environmental organizations are very sophisticated at it. "In the heat of battle, environmental leaders do not think about using the media as much as they should," comments Douglas Foy. "They need to think of the press as a major element in their weaponry."

Who should manage media relations for an environmental organization? Most of our leaders pointed out that public-relations firms are too expensive; but if a group can afford it, a full-time specialist can be extremely helpful. If a group cannot afford a full-time person, part-time consultants are suggested. However, since the majority of groups may not be able to afford any of these specialists, it is suggested that they appoint one staff member to handle this responsibility, and provide this person with training materials and courses.

A point that was stressed by many leaders was that good media relations need to be cultivated. "Don't wait until an event happens to call up the media," suggests Tom Arnold. "Get to know reporters and editors before a crisis arrives so that when you call, they will know you, and you will have credibility." In addition, several people advise being selective. "Don't bury the media with three press releases a week," advises Tom Deans. Brock Evans adds, "Only seek coverage when you have something newsworthy. That builds credibility and guarantees more coverage." To reinforce your relationship with the media, it was recommended that reporters and editors be thanked for providing coverage.

What can a group do to increase or improve media coverage? The ideas offered by our leaders included the following: (1) Select the medium best for your purpose. "The press transfers information and creates understanding, while television creates visual images on which opinions can be formed," observes Gerard Bertrand. The media likes dramatic stories in which there are contestants, winners, and losers. (2) Prepare press releases that are short, simple, and quotable. "Do not get bogged down in jargon or detail," advises Brock Evans. Nancy Anderson comments, "Don't use abbreviations or technical language. Reporters usually won't tell you if they don't understand." (3) Call papers and stations both the night before and the morning of an event. Because there are different people on different shifts, this will increase the chances of contacting a receptive editor or reporter. (4) Send out a release dated on a "dead" newsday, such as a Monday; provide press kits with background materials for reporters. (5) When appearing on television, use graphic materials.

Network Coalition and Building

The final issue discussed with our sample of environmental leaders concerned networks and coalition building. Although strong differences of opinion existed about how effectively environmental groups work together, there was unanimous agreement that cooperation and collaboration are absolutely vital if the environmental movement is to be successful.

The major reasons for developing coalitions and for collaborating are extremely practical, observed a number of people. It is a way of saving money by avoiding duplication, and it creates stronger educational and political impacts than any group can individually. "There are so many issues that none of us can cover all the ground," acknowledges Kelly McClintock. "We need to spread ourselves out as efficiently as possible." In addition to agreeing to have groups focus on different issues or particular aspects of an issue, collaborating and networking can be an economical investment. For example, groups can help one another by sharing office space or equipment; groups can take turns sending someone to a conference and bringing back information to share; materials, equipment, and supplies can be purchased jointly to reduce costs; and groups can combine resources to increase the quantity and improve the quality of training for staff and board members.

Several people observed that, while there are obvious advantages to environmental groups working together, there are many potential obstacles. For example, as Roland Clement points out, "There is often jealousy among groups who are competing for money." "There are a lot of egos tied up in coalitions. Everyone wants his or her name listed first. They all want to make sure that their organization receives the proper amount of credit. It can be touchy to handle," observes Tom Deans. "Once coalitions are set up, you get into difficult questions about who makes the decisions and how they are made and whether they are binding on coalition members," states Tom Arnold.

Despite these problems, all of the leaders we interviewed said it was worthwhile to work at overcoming them. "It is best to compromise to achieve unity," advises Tom Kimball. "Involving oneself with other organizations does not mean any loss of autonomy," reports Kelly McClintock. In a hopeful observation, David Zentner stated what was one of the strongest points of agreement in our survey, "On serious issues, the cooperative spirit will transcend competitiveness."

4

Fund Raising for Environmental Groups

Allen H. Morgan

Fund raising is a matter of survival for any environmental organization. This chapter describes some approaches to fund raising that I have learned from 30 years of experience as director of the Massachusetts Audubon Society and as a fund-raising consultant to other environmental organizations.

Memberships: The Key to Fund Raising

The solicitation of dues-paying members is the single best approach to fund raising. In the long run, the financial and programmatic success of environmental groups is directly related to success in recruiting new members. The majority of the most stable and influential groups, including the National Audubon Society, the Sierra Club, the National Wildlife Federation, the Friends of the Earth, the Appalachian Mountain Club, state and local Audubon Societies, and watershed associations, are supported principally through membership dues.

Developing a membership base does more than provide financial stability. New members become converts to a cause, centers of influence, and the primary political force behind a group's efforts. The idea of being a dues-paying member also has the important connotation of one's belonging to a group whose values one shares. A member makes an unstated intent to continue supporting the organization, as opposed to making a simple one-time gift. In this respect, membership is something you send a bill for; it is the member's ticket to the club, with at least some modest quid pro quo in return. As a result, there is more pressure and incentive to pay dues than to respond to an appeal for gifts.

Another important aspect of developing a membership-based fund-raising program is that, when managed properly, up to 20 percent of a group's membership will make gifts in addition to paying dues. In examining membership and gift statistics in many nonprofit groups, I have found that those groups that have no structured membership program, and simply make appeals periodically, have a rate of giving less than half of that of membership groups.

41

Other Fund-raising Methods

To propose that membership solicitation is the best approach to fund raising for environmental groups is not to argue against other approaches, such as project grant proposals, appeals to foundations and corporations, direct mail, income-producing programs, selling products, and deferred giving. Membership development should be viewed as the core of a diverse fundraising program that can grow over time. Members themselves are invaluable resources in fund raising. They make gifts in addition to their dues; a small number will leave bequests which are a principal source of unrestricted capital funds; and others are key contacts in securing government, foundation, or corporate grants or contracts.

A practical consideration about grants and contracts is that they are substantially influenced by evidence of stability and public support, which are demonstrated through a strong membership base. If you have a large membership, you may have a competitive edge. Another point is not to become too dependent on grants and contracts, because they can arrive and depart suddenly, and thus are an unstable part of a group's finances. They also tend to generate hidden and unanticipated overhead costs.

A great temptation for many environmental and other nonprofit groups today is to turn to business corporations for funds. While this may provide support for some environmental groups, it is no replacement for a good membership program. Remember the simple statistical fact that 90 percent of all philanthropy in the United States comes from individuals. Further, there are serious problems with corporate giving to environmental groups. The decision to give is usually made by a number of people, and a gift to any environmental group worth its salt will be controversial to some corporate executives. A gift implies an association with a group's controversial or unpopular stand, and many corporate leaders may fear that poor publicity could be generated.

Corporate products and equipment are a different matter, and far easier to secure than a cash grant. But regardless of the gift, it is a rare company that will support a group without receiving something in return, such as good publicity or employee benefits. While this may be hazardous to an environmental group's credibility, there are many instances where the benefits are mutual without that risk.

In addition to soliciting from corporations and submitting proposals to foundations and government agencies for grants and contracts, direct mail and deferred giving programs can be very beneficial to environmental organizations. These methods will be discussed in more detail below.

The Importance of Communications

If membership is to be the core of support for an environmental group, then membership recruitment and retention must become a major objective in every aspect of the organization's effort. In its broadest terms, the strategy for building financial support for your organization is straightforward:

1. Stand for something important.
2. Do something about it.
3. Tell your story effectively.
4. Ask as many people as possible for help.

While most environmental groups are good at the first two things, too few are effective in telling their story and asking people for help. Communication is the single most important ingredient in fund raising; if people don't comprehend your need, they won't support you. Communication must encompass the entire range of a group's activity, including media publicity, the group's own newsletter, and even how the organization answers letters and the telephone. An important principle in communication is to *stick to the issues*, because they are the motivation and power behind your work. People will be most interested in these things rather than in internal affairs of your organization.

It is important to recognize that competition for people's attention is severe. Groups must use ingenuity and creativity and be alert for opportunities, such as rebutting media editorials, and stimulating speaking invitations before appropriate audiences, to make their group and concerns known.

Building a good communications effort to reinforce fund raising is slow and tedious. As a rule, I believe that it is better to provide some communication frequently than to send out a lot of information infrequently. Repetition and reinforcement is vital to keep people informed and to keep up with your audience, *which will change by at least 20 percent per year*.

In planning your communications efforts to support membership-oriented fund-raising efforts, remember that you have a number of target audiences:

1. The general public, which is reached through your media efforts, your facilities or properties, and through your programs which people attend. Also direct mail, speaking engagements, and word of mouth through your present members are contacts with the general public.
2. Good prospects for membership are that small segment of the general public who had indicated in some way that they are interested in your organization. The trick in identifying these people is to devise mechanisms to gain their attention and have them identify themselves.
3. Present members are a third audience. Your own internal publications, such as newsletters and special mailings, are, of course, designed primarily for this group. Remember also that your communications effort aimed at the general public will reach and reinforce the convictions of your members, thereby enhancing their support and influence on their friends.

Another important element in fund raising is speaking engagements. Giving a speech is a golden opportunity to tell your story to the public in a highly personal and direct way. Become an effective speaker (take lessons if necessary), and remember the adage that you never get a second chance to make a good first impression. Your personal appearance before groups must be excellent. Your delivery may be improved by using visual aids, such as a dual dissolving projection system which can turn a routine performance into a highly professional presentation at modest cost.

Recruiting New Members

A first step in developing a membership recruitment program is to study what kind of people already belong to your organization and where they live. For example, a membership profile obtained by the Massachusetts Audubon Society a number of years ago found that 80 percent of the members were over 35; 50 percent were over 50; 75 percent lived in the suburbs and most in the outer suburbs; 40 percent had children under 16; and 50 percent held management or professional positions. Membership profiles of other environmental groups show very similar characteristics. This kind of information is critical in helping you decide to whom you should aim your advertising, direct mail, and recruiting efforts.

The very best prospects for membership are those people who have identified themselves by responding to your advertising, phoning with a question, visiting one of your facilities or properties, participating in a program, subscribing to one of your publications, or purchasing one of your products. If you do not have mechanisms for securing the names and addresses of these people, you are missing a golden opportunity.

In addition to these methods, consider using public service announcements on radio or television or writing a newspaper column as ways of attracting new members. While these methods do not raise income directly, they provide a good means of encouraging people to contact your group for additional information. Those who respond are excellent membership prospects.

By all means try to obtain prospects' names from existing members, staff, and board members. Such prospects are usually excellent; although as a practical matter, few of your existing supporters will bother, and the volume from those who do will be small.

Developing membership among young people is a special and difficult problem. Although most children are very interested in nature study until the age of 12 or so, the turnover rate of child members is very high. However, providing camps and special programs for youth is a valuable investment in their future support, and attracts the support of their parents.

Direct-mail solicitation is another valuable means of attracting new members, and is relatively straightforward. Although commerical mail-order companies are of great help and should be used, there are hazards. Be sure to retain control of the operation; establish a good trust relationship with the key company representative; and be wary of signing any contracts that preclude your doing in-house solicitation or requiring that you share those proceeds with the mail-order company.

As far as mailing lists are concerned, your own internally generated list, although smaller in volume than what a mail-order house can provide, is by far the most productive. But such a list will not exist unless you organize to develop it. You should solicit those on your in-house list three or more times, and keep soliciting until the return falls below 1 percent, which is usually considered an acceptable return among commercial direct-mail lists.

One other direct-mail method, which might at first seem like folly, is to enter into a reciprocal arrangement with another environmental organization to solicit the members of each group on behalf of the other. Although you might fear that your members might abandon your group in the process, it was my experience that this procedure never reduced membership and, in fact, resulted in substantial benefit to each of the cooperating groups.

Membership Renewal

Members fail to renew for a variety of reasons, many of which are beyond your control, such as moving away or financial reverses. A sobering fact to remember is that, typically, more than half of membership losses occur at the end of the first year. Treating new members with particular care can reduce such erosion dramatically.

Among the methods that I have found helpful is to promptly acknowledge them by letter or phone (or both!) when they join, and then to follow up in a month or two with a packet of materials for new members. Invite them to a new-member open house: few will come, but the invitation itself is appreciated. Three or four months before their first membership anniversary, send a personal letter expressing appreciation for their membership support, the importance of your work, and the need for continued support, and inviting constructive criticism or suggestions for improvement.

Do not solicit additional gifts from members during the first year. First-year members seldom make additional contributions, in my experience, so little is lost. Additional gifts will come in time as their interest and conviction is heightened by your newsletters and other communications. By using these methods, it is possible to substantially increase the first-year member renewal rate. Using this model, the Massachusetts Audubon Society was able to reduce first-year losses by more than one-third.

The mechanics of membership renewal billing and follow-up will vary among organizations, but all need careful organization to manage what is a very complex and detailed office operation. Managing the renewal process well brings in money sooner and fortifies your image as a well-run organization. Poor renewal management, on the other hand, is a certain route to substantial and unnecessary membership loss. Because of the importance of good membership records and renewal procedures, computers and word processors are becoming a necessity. The sheer competition for support among groups, the need to treat members intelligently and personally, and the high cost of manual record-keeping compel environmental groups to use computerized data processing if they are to continue to prosper.

Another important element of a good membership renewal system is a strategic renewal billing policy. Do not send renewal notices on a calendar-year basis, but rather on the anniversary of the original membership. Since many people do not renew their membership immediately and are forgetful, you must be prepared to send at least three notices at 30-day intervals, the last one being first-class mail with return postage paid. If a member does not renew, do not be afraid to drop him or her from membership, since many people will not pay up until you do this. Of those who still do not renew, a number will join again if you solicit them six months after they are dropped. Reinstatement of former members is often the most productive single source of "new" members.

Keep your renewal appeal focused directly on renewal: do not divert attention by enclosing information about programs or merchandise. Tests conducted by many groups, however, indicate that enclosing a membership card and an auto window decal with the first bill increases renewals. *Do not* send out renewals during the same month you made an appeal for an extra gift.

Upgrading Memberships and Soliciting Additional Gifts

An important element in fund raising is the upgrading of membership dues. This involves two considerations. First, you should establish a number of membership categories in addition to a basic membership. This encourages members to raise their level of annual membership support by voluntarily electing to move into a higher priced category. Second, from time to time you will have to increase your fee structures to deal with the realities of inflation. I have found that it is best to establish as low a price as possible for an entry-level membership, and then to work aggressively to upgrade members. The best time to increase fees is when new memberships and renewals are growing or remain stable. They should be increased to a level that is related to your overall average dues payment.

Your group's most generous supporters are extremely important and deserve special treatment. Recognizing such persons is appropriate for a variety of reasons, including basic equity. While some will shun it, others will increase their support because of the recognition. Holding special programs for these people exposes them to staff and board members, and reinforces their understanding of and interest in the organization. In addition, try sending special mailings to such individuals—perhaps an article, copy of a speech, or a summary of a conference you have attended, with a short covering note.

Because as many as 20 percent of your members will make a contribution in addition to their membership dues, an annual gift solicitation among members is a key ingredient of fund raising. But don't let such gift solicitations compete with your dues bill: send your appeal six months from the date of the anniversary of their membership. This avoids competing with a membership renewal, and the embarrassment of a gift appeal arriving a week or a month or two after the person has paid his or her dues.

Acknowledge all gifts—it is worth the expense. It is part of your communication effort. Try telephoning contributors of $100 or more. Invite larger contributors to lunch (they almost always pick up the tab) to report personally on your efforts and why their support is so important. During such occasions I don't believe it is wise to ask for additional funds unless the donor invites such discussion.

Caring for the Long Term

Beyond annual fund-raising efforts, concern yourself with building long-term sources of income through endowments, memorial gifts, and deferred giving. In accepting gifts of land or other property, ask the donor to match the value of the property in cash to create an endowment fund to support its operation. If the donor cannot or will not do so, ask the board of directors to require that it be raised in a public campaign. In my experience, this is one of the only opportunities to attract endowment gifts from members and the public.

Memorial gifts can also be a source of endowment. While establishing special funds in the name of a benefactor—for programs, buildings, property, rooms or equipment—may often be helpful, avoid it whenever possible, especially for small gifts. Treat such funds as endowments since it adds a sense of permanence to the memorial gift, thereby increasing the amount of the gift or the number of people willing to contribute to the fund.

Deferred giving is the climax to your fund-raising strategy. Basically, deferred giving involves developing an agreement for a future gift which the prospect will not or cannot now afford to make. The best prospects are long-term members and donors. To encourage such deferred gifts, plant the idea in your newsletters, annual report, and anniversary letter. The tenth or twenty-fifth anniversary of a person's membership renewal is an ideal time to contact them about considering a deferred gift. The legal mechanics for establishing these kinds of gifts are technical, complex, and subject to changes in state and federal laws. You will need an attorney to handle the details, while staff and selected board members discuss the idea in more general terms with prospective donors.

The Cost of Fund Raising

It should be acknowledged that fund raising costs money. It requires a major organizational investment of time, effort, and cash if it is to succeed. How much of the strained budget of your organization should be spent on fund raising? Clearly, if fund-raising costs are as high as 90 percent of a group's budget, as was reported in several celebrated cases a few years ago, it is indefensible. Fortunately, in my experience, a realistic figure for fund raising ranges from 15 percent to 30 percent of a group's budget, or slightly higher. But even arriving at such a figure is somewhat arbitrary since there are many activities that may or may not be considered to be a part of fund raising.

The key is to avoid the temptation to slash financial development costs when your budget gets tight. Remember that without a membership constituency, you cannot operate effectively, and you cannot develop or expand that constituency without investing in financial development. Finances are always tight for almost every environmental group, so resist with all your might taking money away from these efforts. The only solution to too little money is to increase your income. To do that you must hire the people and allocate the funds necessary for the survival and growth of your organization.

5 How to Write a Proposal

Stuart Langton

In our complex society, with many organizations competing for funds from several sources, the proposal has become a principal medium for determining who receives how much money for what purposes. Proposal writing has become a necessary skill for leaders of environmental groups.

Perhaps it is the competitive nature of proposal writing, or its growing necessity, or the fact that few of us are trained in how to write a proposal that makes the task appear formidable. Whatever the reasons, many dedicated and capable leaders of environmental groups express considerable uncertainty and have mental blocks about writing proposals. This is unfortunate since proposal writing is not that difficult, mysterious, or esoteric. It seems harder to do than it really is, and it is easier to do each time.

Types of Funding Sources

There are several types of funding sources for which proposals are written. These include:

government agencies (national, state, and local)

national foundations

small family foundations

local businesses

The United Way and other general philanthropic associations

groups or organizations (such as the Rotary Club, Chamber of Commerce, and so on)

individual donors.

In each source, the conditions and requirements for a proposal may vary. Therefore, some initial things to determine before preparing a proposal are:

1. is your group eligible for funding from the potential source?
2. is the kind of project you are proposing relevant to the funding source?

49

3. what is the general level of grants available from the source?
4. what, if any, guidelines, directions, or forms are required?
5. what is the deadline for submitting a proposal?

You may obtain this information by either calling or writing to the funding source. As a rule, government agencies usually have written guidelines and forms, whereas groups, small family foundations, and local businesses seldom have formal guidelines.

Some government agencies or larger foundations may require or request a short preliminary proposal or concept paper prior to accepting a complete proposal. Some small foundations or groups may request a preliminary letter of inquiry with a brief statement of your intended project.

Given these variables, *be sure to carefully check the rules of each procedure before you begin.*

Ten Common Elements of a Proposal

Despite the differences in funding sources and requirements, a good proposal usually contains the following ten elements:

1. a summary of the proposal (brief and to the point)
2. introduction (defining the problem, proposal intent, and terms)
3. need identification (why the project is needed)
4. goals and objectives (what you hope to achieve)
5. methods (how you will achieve objectives)
6. schedule (when you will achieve objectives)
7. evaluation (how you will determine whether you achieved objectives)
8. organizational and staff-capability report (why you are able to undertake the project)
9. budget (what it will cost and why)
10. follow-up and future-funding report (what will happen after the project).

Although you may not be required or may not desire to have your proposal organized into these ten sections, it is important to include all of these items within a proposal. In the following pages, suggestions and illustrations are offered concerning all ten areas.

Getting Started

John Galsworthy, the novelist, once pointed out that "the beginnings and endings of all human undertakings are untidy." This is certainly true with proposal writing because you cannot actually begin to write a proposal until you do some planning. Therefore, I have found that before one even begins to write a proposal, it is helpful to make notes that outline your ideas in four areas:

1. *What You Want to Do* Make a list of the major activities of the program that you will be proposing. Next to each activity, jot down some of the practical tasks associated with each activity.
2. *Budget Estimate* Approximate the cost associated with each activity, as well as staff and overhead costs. This may require obtaining estimates. For example, if you are to use consultants to conduct a study, you should have several consultants provide a general estimate. If you will need office equipment, obtain bids for rental or purchase from several suppliers.
3. *General Schedule* Estimate when activities will begin and how long they will take to complete.
4. *Manpower Assessment* Determine how many and what type of staff will be needed to carry out activities.

One way to organize your thinking in these four areas is to develop a simple one- or two-page worksheet. You can draw up such an outline by yourself or with colleagues.

Assume, for example, that you and several people from surrounding towns want to protect a river that runs through several communities and is a source of water supply and recreation. So, you decide to form a watershed association. You have a meeting and develop some ideas of how you can obtain relevant information about the watershed, educate the public, and persuade the towns to protect and preserve the river. The local regional-planning agency agrees to house and assist the association.

Table 5-1 illustrates a simple method that can be used to plan a proposal for such a project. It contains the heart of your proposal and an outline from which to work.

The Introduction: Some Alternatives

Once you have developed a general worksheet for your proposal, you can begin writing. Your first step is to write an introduction. (Don't begin with the summary—saving that until last will be much easier.)

The introduction will probably be the most difficult, and yet most creative, part of your proposal. The reason is that there is no standard or superior type of introduction, since different types of introductions are written for different purposes. Therefore, the best approach in writing an

Table 5-1
Proposal-Planning Worksheet
Beautiful River Watershed Association

Major Activities	Tasks	Cost Estimate	Schedule Estimates
1. Establish a board of directors	• identify key people in each town • meet with potential members • elect officers • develop organizational plan and bylaws • meet monthly	$500 for meetings	2 months to recruit and organize
2. Conduct a study of the present and future importance of the river and problems in protecting the watershed	• hire consultants • obtain representatives from each town to assist consultants • print final report	$8,000 for consultants $2,000 for printing	9 months to complete study 2 months to edit and print
3. Prepare and distribute materials to educate the public	• distribute consultants' reports to all town officials • prepare brochure for every household • prepare fact sheets for each community	$4,000 for brochures $1,800 for mailing $1,200 for fact sheets	2 months to distribute 1 month to edit, print, and distribute 2 months to edit, print, and distribute
4. Hold a conference to focus attention on the river's importance	• open conference to public • publicize in newspapers, radio and posters	$500 promotion $500 conference cost	3 months to plan
5. Develop an action plan to protect the river	• write plan (staff director) with input from representatives from each city and town • print plan	$1,000 printing	6 months to prepare 1 month to print
6. Keep people informed	• develop newsletter (bimonthly) • develop mailing list • write monthly column in local papers	$300 printing each issue $200 for postage each issue	Can begin by the 6th month

7. Work with each town to preserve and protect the watershed

- meet regularly with each planning board, conservation commission, and board of selectmen
- pass ordinances and articles in each town to protect and preserve the watershed

Notes:

Major Staff Implications: Proposal can be undertaken by one staff director, and a secretary (half-time).

Major Schedule Implications: It will take two to three years to complete project.

Major Budget Implications: Annual staffing and general expenses: $30,000; year-one additional expenses: $10,000; year-two additional expenses: $10,000.

introduction is to remember that this is your first impression—and decide
what type of impression you want to make. For example, three common
types of introductions and types of impressions are:

*The Analytical Introduction—"These People
Really Know What They Are Talking About"*

If you have the ability to present a particularly perceptive analysis of the
general problem area of your proposed project, you may want to demonstrate your particular grasp of the issue you are addressing. In such an instance, the introduction can be more analytical and reflective; however, it
should not be excessively academic or lengthy.

*The Organizational-Capability Introduction—
"These People Certainly Have the Credentials"*

If your organization and/or proposed staff have exceptional experience or
background, then you may want to stress your credentials at the outset.

*The Crisis-Problem Introduction—"That's a
Pretty Serious Problem"*

If the problem you are addressing is particularly severe or dramatic, you
may want to stress its intensity or significance. In such cases be sure not to
overstate the case, as you may develop it more fully in a following section
on needs. However, you may want to combine the sections and entitle the
first section, Introduction: The Problem and Needs.

   Once you have decided upon the general impression you want to make
in an introduction, the next thing that is particularly helpful is to develop a
good opening line that is consistent with the impression you wish to make.
So, for example, the watershed association proposal might begin quite differently:

Analytical Introduction. "Water supply is one of the most critical but
underestimated natural resource problems in the eastern United States. One
reason for this is. . . ."

Organizational-Capability Introduction. "In a recent meeting sponsored
by the ad hoc Beautiful River Watershed Group which brought together
Governor Zippo, U.S. Senator Sam Straight, selectmen and mayors from

fifteen towns, and representatives from seventy-five citizen organizations, it was unanimously agreed that a Beautiful River Watershed Association should be established.''

Crisis-Problem Introduction. ''On May 15, 1983, the City of Zombie was forced to discontinue using the Beautiful River for its water supply when 178 persons became ill as a result of hazardous waste contamination of the river.''

Problem and Need Identification

Every proposal should *clearly define* what the problem is that you are addressing. Then you should *document* why the problem is important or why there is a need to address it. In terms of identifying and documenting needs, there are three things that are particularly helpful: First, draw upon research and statistics to document the problem that you claim exists. Second, quote authorities or those who are close to the problem to describe it more thoroughly. Third, indicate why the problem is not being addressed adequately and describe the consequences of our not addressing it.

One bit of advice concerning the use of statistics: don't overkill. If you have a lot of charts, illustrations, and lists, put them in an appendix. Use only your clearest and most dramatic proof in this section. Put additional supporting material in the appendix.

Goals and Objectives

In this section of the proposal, you will be describing what you hope to achieve and the things that must be accomplished to achieve it. The term that is used to describe the final achievements or outcomes is a *goal*. A step or accomplishment that is necessary to achieve a goal is an *objective*.

For example, in the illustration of the Beautiful River Watershed Association, a goal would be ''to preserve the flood plain of the Beautiful River in its natural state.'' Among some of the necessary objectives or steps in achieving this goal would be:

1. To prepare a map of the entire Beautiful River basin that clearly identifies the flood plain.
2. To directly inform all elected and appointed officials and leaders of citizen organizations in communities adjacent to the river of the importance of flood-plain preservation.

3. To prepare a written report that clearly identifies the flood-plain marsh-
    land area in every community in the Beautiful River Basin and the
    owners of the marsh and land in the flood plain.

In preparing objectives, here are a few tips: First, begin each objective
with the word *to* and follow it with a concrete action (that is "to *prepare* a
map"). Second, objectives should be so clear that it is possible to observe or
measure if they have been achieved. Always ask yourself: Will we be able to
see or measure the achievement of this objective? Third, don't confuse your
objectives with your methods. An objective is an accomplishment, but a
method is *how* you will accomplish the objective. Further, it may require
several methods to achieve an objective. As an example, consider objective
Number One above, "to prepare a map." To do this may require several
methods such as: (a) obtaining existing maps from the U.S. Army Corps of
Engineers and other agencies; (b) taking aerial photographs of the flood-
plain area at the high-water season, and (c) hiring a map maker and graphics
consultant to prepare smaller scale copies of a master map, and the like.

Methods and/or Activities

Within your proposal you may choose to list the various methods or activities that you plan to undertake to achieve your objectives. Each description of a method or major activity should contain some detail of the
steps and conditions involved in carrying out the method or activity. As an
example, assume that you identify some principal activities to carry out the
Beautiful River Watershed Association proposal, including the following:

Example of a Method/Activity

To serve the goals of this project (establishment of a board of directors), it
will be necessary to create an influential and respected group of citizens to
establish policy for the proposed association and to assist in education and
advocacy efforts in the communities in the Beautiful River basin. The board
will consist of one representative from each of the fifteen cities and towns
within the basin area. The League of Women Voters of Greater Zombie will
obtain the names of five to ten interested and respected potential representatives from each community. The board of selectmen or city council of
each community will be asked to select a representative from the list of
names submitted to them. Once this process is completed, the board will
meet for a weekend work conference to elect officers and draft bylaws. It is
intended that this process be completed by the third month of the project.

Some hints in describing methods: First, describe how the method or activity will be carried out, but don't go into too much detail—just give the essentials. Second, provide a title for each activity such as: establishment of a board of directors; map study; newsletter; and so on. Third, as possible, indicate when an activity or method will begin, how long it will take, and when it will be completed.

Schedule

A schedule is simply a graphic means of indicating when each project activity begins and ends. As such, it supplements the earlier description of methods and activities and provides a visual summary of intended project progress.

While there are many types of schedules that can be used, I have found that a simple one-page milestone chart (such as the one contained in table 5–2) is the simplest and clearest.

Evaluation

In developing an evaluation plan for a proposed project, it is helpful to make a distinction between (ongoing) process evaluation and terminal evaluation at the end of the project. The principal purpose of the former is to provide feedback that the staff can use in planning. The principal purpose of the latter is to indicate the extent to which each project objective has been achieved.

Evaluation activities can be undertaken by project staff as well as by an external consultant. As a rule, it is appropriate for purposes of objectivity to have a consultant conduct the terminal evaluation. In planning a proposal, the following steps may be helpful in preparing an evaluation plan. First, have someone who is experienced in evaluation review each objective to determine the most simple and least costly way of determining the achievement of each objective. Second, identify an evaluation method for each objective. The most common evaluation methods include questionnaires, interviews, and observations. Third, determine and identify which evaluation method can be undertaken by staff to facilitate process evaluation and which methods will be undertaken by a consultant as a part of the terminal evaluation. Include plans for the external consultant to be involved from the outset of the project. Fourth, select a consultant and obtain an agreed-upon budget for preparing the evaluation. (As a general rule, it should not exceed five percent of the total cost of the proposed budget.) Include in the proposal the name of the consultant, be it an individual or an organization, and include a resume or an organizational-capability statement in the appendix.

Table 5-2
Milestone Chart, 1984-1985,
Beautiful River Watershed Association Proposal

Major Activity	Jan.	Feb.	Mar.	Apr.	May	June	July	Aug.	Sept.	Oct.	Nov.	Dec.
1. Establish board of directors of Beautiful River Watershed Association	O		▲									
2. Create watershed map	O				▲							
3. Study present and future importance of river and problems		O						▲				

Notes:
. . . etc. (Listing of remainder of fifteen major activities) . . . . .
O - indicates commencement of activity
▲ - indicates completion of activity

Organizational Capability and Staff

In this section you should include a description of the background of your organization and its major accomplishments. In particular, you should indicate why your organization is particularly suited to undertake the proposed project.

You should also identify the staff who will undertake the project. Identify each staff position, the functions the staff person will perform, and the background of persons for each position. If you want support for a new staff person through receiving the grant, you should include a job description. If the staff person is already selected, include a brief description of his or her background, and include a copy of their resume in the appendix.

Budget

A good budget usually contains two sections. The first is an itemized list of proposed expenses such as that appearing in table 5–3. If the project proposed is to continue more than one year, the budget should include a column for each year. In this example, you could begin by stating that the total cost of the proposed project would be $91,971, of which $53,020 would be needed in the first year and $38,951 in the second.

Following the budget summary, you may want to add a section entitled Budget Narrative. In that section you would briefly explain the proposed expenditures for each item.

Plans for Follow-Up/Continuation

Funding sources are usually interested in seeing that their investment in your proposed project will generate an effort that will be continued. Therefore, it is appropriate to indicate how the project will be continued after the initial funding period. In the example of the Beautiful River Watershed Association, you could outline a plan for funding a project budget of approximately forty thousand dollars after the initial funding has expired.

An additional thought: If you feel that the project is a one-shot deal, say so, and offer arguments for this. However, if you feel the project should be continued, describe a realistic game plan for continuation.

Project Summary

Once you have written your proposal, you can write the summary. The summary, which will be placed at the beginning of your proposal, should include

Table 5–3
Proposed Budget Summary:
Beautiful River Watershed Association Proposal

	Year 1	Year 2
1. Staff:		
Project director (full-time)	$20,000	$22,000
Secretary (half-time)	4,000	4,400
2. Fringe benefits (15%)	3,600	3,960
3. Consultants:		
a. to conduct study ($10,000)	10,000	
b. to prepare maps ($1,500)	1,500	
c. to serve as conference resources		500
4. Printing:		
a. map	500	
b. study	2,000	
c. action plan	200	
d. stationery	300	
e. newsletter	2,000	
5. Office equipment (rental)	1,000	1,000
6. Office supplies	500	500
7. Telephone	1,000	1,100
8. Postage:		
a. general	200	250
b. report distribution	200	
c. newsletter	300	900
9. Travel:		
a. staff	100	500
b. consultants	500	
10. Miscellaneous	300	300
	$48,200	$35,410
Overhead: 10% of project cost	4,820	3,541
	$53,020	$38,951

the highlights of each section of the proposal. The summary should be short—one, or no more than two, pages. Whenever possible, quantify items ("The newsletter will be distributed to three thousand persons in year one and eight thousand in year two." "The conference will seek to attract five hundred interested government officials and interested citizens," and so on).

The Appendix

An appendix includes supporting material that goes into greater detail than the main body of the proposal. An appendix may include such materials as:

1. additional documentation of the problem addressed by the proposal
2. background information about your organization, your board of directors, and staff (possibly including resumes)
3. examples of materials you may use (if applicable)
4. letters of support.

Some Final Thoughts

Here are some final miscellaneous suggestions in putting your proposal together:

1. Letters of support. It is often helpful to have letters from other groups and government officials that indicate interest in and support of the proposal. If you do plan to include supporting letters, be sure to request them as early as possible, telling the person what you want said and providing a deadline. You may even want to send them your working outline. Keep a list of the people you ask and a folder for the return letters. Call those who haven't returned letters a week or two before the deadline.

2. Cover Page. The cover page of your proposal should include: the proposal title; the name of the organization to which you are submitting the proposal; the date of submission; the title, address, and phone number of your organization; and the name of a contact person in your group who is familiar with the content of the proposal.

3. Cover Letter. A cover letter should include a simple statement of what you want to do and why the proposal is important to your group. It should be gracious, and it should also identify the person to contact in case of questions.

4. Copies. Make arrangements to have clear and multiple copies made of the proposal. Many funding sources request multiple copies. Even if there is no such requirement, send two copies, as one may get lost or temporarily misplaced. Be sure, also, to make a sufficient number of copies for your own organizational use.

5. Delivering Proposals. If you are submitting a proposal to a government agency, always send it by registered mail (return receipt requested). If it is delivered in person, obtain a receipt with the date and time delivered.

6. Calling the Funding Source. Don't be reluctant to call your funding source. It is perfectly all right to call while you are writing the proposal or after you have submitted it to ask a question.

7. Length. Grants are made on the quality of a proposal, not on their length. It is important that you know what you are talking about, and you say it as simply and clearly as possible. Avoid jargon and excessive verbiage.

8. Style. Obviously, clarity and good grammar are never out of style and are always appropriate. In addition, organizing your text into sections and subsections with appropriate titles is a good stylistic procedure.

9. Graphics. While a picture may be worth a thousand words, a graphic illustration may be worth five hundred. As appropriate, develop charts or graphic illustrations to summarize and illustrate important points of your proposal.

10. Outcome. You may find it helpful to know that the majority of proposals are *not* funded. For example, many government grant programs fund only one proposal for every fifty or one hundred submitted. Therefore, you should consider submitting your proposal to a number of carefully selected sources, and recognize that you may want to alter parts of it to be particularly responsive to these different funding sources. Also, from sources that do not fund it, you may ask for suggestions on how the proposal can be improved or where it is deficient. Unless your proposal is a real loser, improve it. And keep trying.

6 Environmental Community Organizing

Lois Marie Gibbs

There are many ways to take action to address environmental issues. At the heart of the environmental movement, however, is the capacity to organize at the community level. Based on my own experience in helping to organize homeowners at Love Canal, and later in other communities across the country, I am convinced that organized grass-roots citizen action is the most effective way to get action taken to address local environmental problems. Because of the effective action of citizens at Love Canal, over one thousand families were able to evacuate the neighborhood with their homes purchased at fair market value. In Enfield, Connecticut, citizens organized to halt construction of an unsuitable waste-treatment plant; and in Riverside, California, citizens fought successfully to have government clean up an illegal and hazardous dump site. Without local citizens becoming actively involved in all three of these cases, the outcomes would have been very different.

While most of my direct experience in environmental community organizing has been related to hazardous-waste problems, I believe that the lessons learned from these campaigns are particularly applicable to many critical environmental problems that arise at the community level. In this chapter I will draw upon recent practical experiences in addressing problems of hazardous waste in order to suggest strategies and methods for grass-roots community organizing over environmental issues.

Finding Out What Is Going On

So, you believe there may be an environmental problem in your backyard? Your children are sick, the air smells funny or awful, your neighbors have told you that they have seen barrels or dumping a few blocks away from your house. Or perhaps a corporation is eyeing your neighborhood as a perfect location for a new hazardous-waste facility, or you have heard that a highway is being planned through the swamp down the street from your house. You need to know how to find out what is going on.

There are many ways to obtain information about such things. The first places you should look are the local planning board, city hall, county health department, and in your local newspaper's library. All of these places may have information that could help you obtain necessary background information. The local planning board, for example, may have specific plans on file. At city hall, you can usually find out who owns the property and whether any complaints have been filed. Your local county health department can also tell you if any health or environmental complaints have been filed, if any environmental tests have been taken and what the results are, and if any future testing or other activities are planned.

The library of the local newspaper may have relevant news articles on file that you may have missed or that were printed before you moved into the area. The state and federal government should have information on permits, permit applications, and environmental testing that might have been done.

The next place you should go is to *all* of your elected representatives at the city, state, and federal levels. These officials can provide key information as well as the names of other people you should contact. There are two things you will accomplish in contacting these public officials: (1) you will obtain valuable information, and (2) you will let them know that you are concerned and, therefore, that *they* had better become concerned. The reason for contacting all the representatives is that each has different contacts within the government and can therefore provide different pieces of information.

Be sure to express to everyone that you expect an answer within a designated time period (two weeks). Then follow up the conversations with a letter saying "as per our telephone conversation on (*date*), I will expect your response to the questions I have raised by (*two weeks later*)." By doing so, you will have a record of your conversation, and you will remind them that they agreed to contact you and provide certain information within a certain amount of time. The letter shows that you are serious, and you now have a record of who said what, and who promised what. This can be very helpful at a later date. Citizens in one state, for example, dealt with a public official who claimed to know little about a proposed hazardous-waste facility. The citizens later learned that this same official owned the property where the new facility was going to be located! Be sure to follow up on those representatives who do not get back to you within two weeks.

Overall, how useful is it to spend time gathering this information and writing the letters? The answer is simple—it's priceless! Love Canal is a good example. Since Love Canal resembled an open, harmless field, most people were unaware that a dump site was located in their neighborhood. After contacting the local- and state-government offices, the residents discovered that Hooker Chemical Corporation at one time owned the property and had disposed of approximately twenty thousand tons of chemical wastes at the site. They also found out that city hall had numerous citizen complaints on file, and that an environmental assessment was conducted in 1976 by the local health officials in cooperation with the U.S. Environmental Protection Agency (EPA). Furthermore, the local newspaper had run a number of articles on the canal and its problems, and the board of education had maintained minutes of their meeting when the land transferred hands from Hooker to the board of education, as well as having kept records describing difficulties the board had in constructing the school building.

If you are unable to obtain information from the agencies and authorities you have contacted by letter or phone, do not give up! You can file a Freedom of Information request. This can be done by simply writing a letter to the head of an agency saying, "Dear Mr. or Ms.: I am requesting under the Freedom of Information Act all correspondence, memorandas, reports, and soil-, air-, and water-test results and analysis, . . . " Be sure to include your name, address, and phone number on the letter in case the agency has a question. Under the Freedom of Information Act, they must give you the information requested within a designated time (in New York State it is ten days).

Having gathered the information, what next? You are outraged at what you have found and you want to do something about it. Your family's health, their future, and their lives may be at stake. Your biggest investment, your home, is devalued, and you are unsure of what to do. You must first realize that you cannot fight this battle alone. As one person said to me, "I went to my authorities alone to ask for help and I was told to go home, to stop causing trouble. I was quickly dismissed as the hysterical housewife." When this person later returned to that same office with hundreds of people behind her, the response was, "Let's sit down and talk."

Organizing a community around toxic waste or similar environmental issues is not as hard as you might think because everyone is affected, whether they care to believe it or not. Their health, their drinking water, the air they breathe and ground they walk on, and the quality of community life can be threatened. Once you show people how they are affected, they are much more willing to become involved.

Initial Steps in Organizing

The first step to organizing is educating the community about the problem. There are several ways to accomplish this. First, call your local newspaper and tell the environmental reporter what you have found out and how outraged you are about the situation. Invite the reporter to look over the documents you have collected and offer to meet with him or her. It would be extremely helpful as a first step toward educating the community if you could interest the reporter in writing a story for the local paper.

The next step is to prepare a fact sheet to be distributed door-to-door. The fact sheet should be kept very simple and no longer than one page. The information might include the following, for example, if you are dealing with a hazardous-waste site: your name, address, telephone number; the exact location of the site or proposed site; what is buried there (if known); what health effects could result from exposure to the wastes; who owns the site; who disposed of their wastes there; and any other information you think the community should know. If you make the fact sheet too long, people won't read it, and if it has too many long names or confusing facts, people will not understand it. Remember, you know much more about the subject than the people you're sending the flyer to; don't assume they know anything.

The fact sheets should be distributed no more than a week before you begin to go door-to-door. Most people have short memories, so you want to be sure they have the issue fresh in their minds when you approach them. The fact sheet can be distributed by you or by other adults, or teenagers, in the community. You might even approach the local Scout troops, where children who need to earn a community-service badge may be willing to help.

You should now begin your door-to-door campaign. The face-to-face contact with the community is extremely important—you will learn more about the community in the process, and residents will ask you questions in order to understand the problem. These people can provide you with valuable information on the history of the problem or what they have seen. When you offer to provide them with answers that you may not yet have, this will give you a reason to recontact them and to begin to build a reputation as someone the residents can turn to who will help them find the answers. Soon you will gain their trust and their confidence.

Before going door-to-door, put together a petition that everyone can sign. The wording of the petition should be thorough but to the point. For example: "We, the citizens of (*your town*), want the *CHEMIKILL* landfill investigated to determine if it is leaking, and if so, what is the extent of the contamination. If contamination exists, we want the site cleaned up to protect public health and our environment." If you circulate a petition making more specific demands, you may alienate a portion of your community who do not understand the circumstances well enough to support you. For example, if you ask to have everyone evacuated from the area, you will find that at this stage of the fight people are not convinced that a severe enough problem exists to warrant such a drastic measure. Obviously, many people will not want to leave. This may occur even when a severe problem exists. These families might not sign your petition or even talk with you. As another example, if you ask the "Chemikill" Company to clean up the site immediately, without all the available information to support you, people working in that industry will not sign the petition because they will fear losing their jobs.

The purpose of the petition is to gain community support and to further educate the residents. Collecting names on the petition can also be used as a political pressure point; public officials will respond to a petition with five hundred or a thousand names on it. They won't respond to one citizen working alone. The petition itself, however, has limited value in actually accomplishing anything. No one in government or public office will look at what it says—only how many voters signed it.

When going door-to-door, you should take a composition notebook, which will come in handy when people give you information that you want to keep. You will be amazed at how valuable this information becomes later on. The notebook (a composition notebook is best so you are not as apt to tear out pages) should always be carried with you, not just when going door-to-door. Remember the data, and write down who told you what.

Before you start going to houses, think about what you want to say to people when you knock on their doors. You can practice your little speech in front of a mirror if you are shy. Knocking on strange doors is always a bit scary at first, but once you have done a few homes, it becomes easier. The thoughts going through your head are most likely not much different than others who have gone door-to-door: Will they slam the door in my face? Will they think that I am crazy or am causing trouble? Will they think I am looking for publicity or going to run for office? Or, rather, will they think that I'm trying to shut down the local industry, thus taking the bread and butter off their table? These are common fears. In fact, I was so frightened that I literally ran away from the first door I approached at Love Canal.

Once you begin your door-to-door campaign, be sure to bring your notebook, the petition, and copies of the fact sheet you have prepared. The following is the way everything should be presented:

1. Knock on the first door, give your short talk about the problem, and ask the resident to sign the petition.
2. Ask the resident if he or she knows anything about the problem or knows anyone who could help you gather additional information (write it down in your notebook).
3. Ask the person if he or she has any questions you might ask the authorities next time you speak with them (if so, write it down along with the resident's name, and be sure to get back to that person even if the answer to the question is, "We don't know").
4. Ask the person if he or she might be interested in helping you knock on doors or doing some other task such as typing or phone calling.
5. Thank the resident for his or her time and explain that you will be planning a meeting in the near future and will notify him or her of the time and place.

Unfortunately, not all the doors you knock on will be quite that easy. A few people will not even want to talk with you, let alone sign your petition. Some may want to argue about the problem or call you names. When this happens, politely thank them for their time and walk away. Never argue with them or insult them, even though you might want to; this can only make the situation more difficult later. A resident who may not at first understand the situation may later support you, but a bad memory of your visit will make this more difficult. It's better to kill them with kindness. I used this tactic at Love Canal. Many who at first did not want to support the organization later became some of the hardest workers. There was a running joke in our office that when the "Gibbs Haters" called, the staff always knew it because I would end the conversation with the words, "everyone is entitled to his or her own opinion and you have certainly expressed yours, so I'd just like to say, have a nice day and do call back again." By being polite, the "troublemaker" could never say, "Lois Gibbs called me a . . ." or "When I called the office she was rude to me; I'm a resident too."

Setting Up a Meeting

After circulating your petition and finding a few volunteers, you are ready for the next step—setting up a community meeting. You must first choose the location. Look around your neighborhood and see what facilities are available. Remember that the closer you are to the residents, the better the attendance will be. Think about walking distance and bus routes; maybe you can even set up a car pool for the elderly or handicapped. The best places to hold a meeting are at a church, library, school, or even city hall because these places generally do not charge rental fees.

Once you have identified a place, pick an evening (most people are busy during the day) when there are few activities elsewhere in your community. Mondays through Thursdays are usually the best days. You might want to schedule a meeting to coincide with the next city-hall meeting. You could, for example, hold your meeting an hour or so before a city council meeting at a room in the same building. In this way, if your organization wants to approach the city councillors for help, it will be easy enough to do.

The next step in planning a meeting is to prepare an agenda. The agenda could include introducing yourself and providing background on the problem, opening the floor for discussion of the problems and issues, and asking whether the participants would like to formally establish an organization (united we stand, divided we fall). If so, the organization should then elect officers and develop committees. Finally, and possibly most important, the group must establish the goals that the members have decided to work toward.

Once your agenda is together and the time and place of the meeting confirmed, you should send out another flyer announcing the meeting. Remember to list where and when the meeting will be held, who is holding the meeting, and what the meeting is about (you might want to include a tentative agenda and ask for other suggestions). Finish the flyer with some strong reasons why people should attend.

The Meeting

The meeting date arrives, and there you are standing in a room with one hundred people (don't be disappointed by a small turnout at your first meeting; interest will grow later). Instead of becoming instantly afraid and sick, remember these are your friends, they are not going to attack you, and that your family's health may be at stake!

Reintroduce yourself to the group. Tell them what you and others (volunteers, whom you should name) have accomplished thus far. It is useful to reiterate the problem even though you or someone else has knocked on everyone's door and explained it once. There will be some people who were not at home or who have not read about the problem in the newspaper. This background information is also necessary to bring everyone to the same level of understanding, thus encouraging participation at the meeting.

Explain to the group that it is much easier to resolve a problem if they band together to form an organization. Explain how you and a few others have done all you can as individuals and how you need *their* help and ideas in order to work effectively in fighting the problem. The "united we stand, divided we fall" ethic cannot be emphasized enough. Tell this to people repeatedly until it begins to sink in. Explain who you are up against: the state, the federal government, or big businesses who have money and power.

If establishing a formal organization is agreed upon by the group, you should publicly elect officers. This can be done by using volunteers to write down nominations and count hands from the audience. If you are running for office, be sure to step aside during this part of the election to avoid any conflict of interest. Otherwise, if the fight gets dirty later on, people who dislike you could try to use it against you. Developing committees at this meeting is also wise, but I would not advise electing chairpeople at this time. Wait until the next meeting so that you can work with people and see who will stick it out and who is best for each committee. You might form specific committees for research, fund raising, action (planning demonstrations), speaking, law and bylaws, and publicity.

It is wise to select one or more spokespersons for the organization. This will keep others from giving press statements (on behalf of the group) that may not reflect the group's opinion. People should be told that when they speak to the press they should be sure to say they are speaking as a victim or as an individual, and not on behalf of the group. A spokesperson should be someone who is constantly involved with the day-to-day workings of the group so that he or she is aware of all developments.

The most important part of your meeting is setting goals to help focus your efforts. What do you as a community want to accomplish? The group should pick realistic objectives; it is usually unwise to have more than three or four main goals. Any remaining aims should be placed on a "laundry list" of additional plans. Instead of making demands, it is wise to state what you want as "the needs of the community." A demand can be argued against; it is also a negative word most often used by spoiled brats. A need, on the other hand, cannot be as easily criticized; it also sounds much more pressing.

Be sure your goals do not contradict each other. I worked with one group, for example, that was fighting against the placement of a landfill in their community. Their goals stated that they did not want the dump in their area; but if it were placed there, they said, it must contain a clay liner. These conflicting goals were interpreted by the authorities to mean that the community was willing to compromise. Because their goals contradicted each other, they lost their fight altogether. Not only was the landfill sited in their neighborhood, but they also lost the clay liner. Their organization was split on the issues. One half fought for the liner; the other half fought to prevent the siting of the facility. There was no cohesive voice representing the entire community, making it easy for the opposition to do what it wanted. In both Enfield, Connecticut, and Canada, however, where the communities organized and together said, "no landfill," the industry could not fight the groups and left town.

Last, but equally important, you must find a name for your newly formed group. Be very careful that the name does not alienate or eliminate any portion of the community. At Love Canal, our organization's name was Love Canal Homeowners Association. This was a misnomer because people who rented homes or apartments felt they were left out even though this was not intended.

When opening the meeting to discussion, guide the conversation by posing questions such as: What should be done? Who are our target figures? What should our first action be? Do we want to go to city hall to ask the city councillors for help? In discussing the targets for your fight, you should be careful where you put your energies. In the case of an existing dump, for example, you need to target the industry—but chances are that the only way you'll receive compensation from them is through lawsuits that are years away. Therefore, you might also want to target the government, banks, or other institutions that can provide you with immediate assistance. Go after the industry, but put more energy in the institutions that can provide you with what you need immediately.

Be sure to have people sign their names, addresses, and phone numbers, either when they enter the meeting or during the meeting on a pad you will pass around. You will need this information in order to get in touch with them later. You should have extra pads available to sign up people interested in volunteering their time or resources. Leave a place on the sign-up sheet where the volunteer can indicate what he or she prefers to do or can do (typing, phone calling, or whatever).

Before you close the meeting, set the time and date for the next meeting, and remind the citizens of what actions the group has decided to take before then. Encourage people to support these actions in such ways as speaking out or writing letters.

This might also be a good time to ask for donations from the group to pay for flyers, postage, long-distance phone calls, and the like. Be careful, however, not to ask for donations at every meeting. People will begin to think you want them to attend only for their money. Even worse, some families may not attend because they are embarrassed by the fact that they simply cannot afford to donate.

The Fight

There are many approaches—including the scientific, the political, and the legal—that can be used when fighting a local environmental problem. Although each of these approaches is useful, no one method alone is good for every situation. As a matter of fact, most groups find that combining all approaches is better than using just one strategy. Following are a number of issues to consider in regard to each of these approaches.

The Scientific Approach

For example, to fight a toxic-waste problem on the scientific front is generally useful but specifically very limited. Assume that you find that the existing landfill or proposed facility is scientifically proven to be a problem. The soil structure is too sandy, which will allow wastes to move readily through the ground—or chemicals have been identified in your drinking water—and a connection to the existing site has been made. In these cases, you have a good piece of ammunition to fight the decision makers. On the other hand, for every scientist you hire that says the site is unsuitable and a hazard, your opponents will find five that say it's safe. They have the money and power to hire these scientists. It requires extensive funds to fight the battle on the scientific front, money that community organizations usually do not have.

Also important is the limitation of scientific understanding. Scientists do not know what will happen to a family's health when the unborn and small children are exposed twenty-four hours a day to chemical soups consisting of many different chemicals. Over two hundred chemicals, for example, have been found at Love Canal. Although scientists may know what one chemical can do, such as benzene causing leukemia, they do not know what benzene, chloroform, and toluene together will do.

Scientific evidence, therefore, can only be used as a part of the overall strategy. In practice, this is where it fits best. At Love Canal, science data was available. It was documented that reproductive disorders were increased in women residents, but the authorities sought only to evacuate pregnant women and children under two years of age! This was irrational and unacceptable to us. In this case, our data was used to argue that "if pregnant women and small children are affected, then we all are in danger." In this way, scientific knowledge was used as a powerful tool to express our position and to gain public support.

The Legal Approach

The legal approach is another way to fight a battle, but is also very costly. In many cases, however, you can find an attorney who will work on a contingency basis. This means that the attorney will take a portion (usually one-third) of the settlement if the case is won. Retaining an attorney in this way helps to ensure proper representation, since he or she does not get paid unless you do. Unfortunately, though, it is next to impossible to hire an attorney under a contingency agreement in the case of siting a new facility because there are little or no damages to collect on.

One of the biggest drawbacks to using the legal approach is that the community will not become involved when legal counsel is brought in—people tend to sit back, thinking the lawyer will take care of the problem. Thus, it is difficult to get people out of their homes to attend meetings.

In addition to these problems, most attorneys are very conservative and want to control the group. Generally, they do not like protests or any other public demonstration, the lack of which might seriously jeopardize the effectiveness of the group. For example, one attorney recommended that an organization not picket a site where remedial construction was proposed. He claimed that the pickets might turn off the public, instead of stopping the work as the community wanted. The community decided to picket anyway. This did not stop the work (the attorney was correct), but the action did force the authorities to give the group many concessions. A consultant was chosen by the community and paid by the government for two years at two hundred dollars a day; further testing was done of the soil, air, and groundwater; health testing of residents was initiated, with a two-hundred-thousand-dollar human-services fund to pay for residents' medical tests; and an immense amount of public support materialized because the public now had a better understanding of the problem from the news stories carried about the protest.

Keep in mind that when you file a lawsuit it will greatly limit your access to information. The government or industry now has the excuse that they cannot give out that information because of pending litigation. Therefore, you should get as much information as possible before you file a suit, as well as securing an agreement from any institution that when further testing begins your organization will receive the results of all tests. On the other hand, if you can keep the above problems in mind and under control, a lawsuit is, at times, very useful in applying pressure. There are many legal avenues open to organizations, ranging from a lawsuit on a local zoning ordinance to application of federal laws such as the Clean Air or Clean Water Act. You should take advantage of these laws in resolving your problems, but be cautious not to eliminate other useful approaches.

The Political Approach

This method is very useful when it is applicable—close to an upcoming election, for example. All elected officials want to look good in the eyes of their voters, so use this knowledge to benefit your group. One way to do this is to hold a public meeting and ask the elected officials present if they support their constituency in (*whatever your goal is*). Very rarely will they say they don't support their constituency—after all, that's who put them in office. If the representative supports you, ask what he or she is going to do and how long it will take (be sure to give a deadline).

If the elected official sidesteps or ignores you, you have to show him or her how it will be beneficial to support you. This can be achieved through a letter-writing or telephone campaign or by attending the official's rallies, dinners, and other public appearances and asking, "Mr./Ms. Representative, do you support your constituency in (*your town*) in cleaning up the *CHEMIKILL* landfill?" If the answer is, "Yes, but we must conduct further tests first," your response should be, "Your constituency does not need further tests; we have enough evidence to show contamination. Do you support us in getting the site cleaned up?"

Have people from your group attend every public appearance and ask the same questions. Believe me, representatives will get tired of your presence and will do something just to appease you. Remember, even if industry puts up the money for the political campaign, officials still need votes and must protect their public image as people who care about the health and welfare of others.

You might consider holding a debate (on the issues) in your neighborhood with the two or three candidates. Use a chalkboard up front to keep score, and only accept yes or no answers. If the candidates answer "I am not sure," or "I can't answer that question with a yes or no answer," mark their answer *no*. Let them all know the rules before the debate, but never give them the questions.

Everything at Love Canal was fought on a political front. Governor Hugh Carey evacuated residents in 1978 only because our organization made him accountable not only in the eyes of our community, but to the whole state of New York. The group put him in front of the public three months before election and asked him if he were going to allow everyone over two years of age to stay in Love Canal and die. He had to answer no—and he evacuated all 239 families. The media followed him around and made his answers known throughout the state. After all, who would vote for someone who could allow people to suffer or die?

Elective politics are very tricky. It is usually best not to side with any one candidate, as you might alienate a portion of your support group. The best way to support a good candidate without making a public commitment is to mention the names of those who have helped you to the media or at public meetings and rallies. Publicly present the candidate with a home-made award for one year of supportive service to your organization. Also, stage a protest or release statements on how badly the alternate party has treated your group. These actions support the candidate, but accomplish this in a way that will not eliminate others in the community who might not like the candidate.

Troubleshooting

Are the scientists and politicians baffling you by their jargon? Often, both the scientific community and politicians will talk in a way that misleads or confuses you to the point that you do not understand what they are saying. The best way to avoid this is to talk with these people on *your* own level and force *them* to that level. You must remember that they spent years of schooling to learn this language and there is no way, even after working with the data and reports, you can compete with them. A productive technique in handling this type of problem is to act as if you were a five-year-old and focus on specifics. Ask "Why? and What if?" Tell them if you don't understand. If you have ever had a small child ask you "why" ten times in a row, you'll know how effective this can be. For example: Q: Why are you only evacuating pregnant women and children under two? A: Because when children reach two, their bodies are completely developed and, thus, can filter chemicals. Q: What if the child is a slow developer? A: Then their bodies may not be capable of filtering the chemicals. Q: Why then are you using two as the cutoff age? What about those children who are slower in developing? A: We have to use some age as the cutoff age; we can't evacuate everyone. The last statement is the one you should look for. It is clear that they are more concerned with costs than the safety of the children, and that the decision was not based on science, but rather on saving money.

The important part of this questioning is not to give up. You may feel funny or embarrassed when you ask "why" fourteen times successively, but this is sometimes the only way you can get a straight comprehensible answer. Also, if someone is trying to hide something from the group, once he or she begins to talk at your level it is much more probable that they will slip and reveal something. Keep in mind, though, that not all scientists who have difficulty speaking with laypeople do it intentionally. Many scientists truly have a hard time turning technical information into language everyone can understand. Try to remember this so as not to alienate a good person who would like to help.

Splinter Groups

In every organization there are a few people who believe everything the group does is wrong. Nothing ever seems to satisfy these people, and they will express their dissatisfaction to anyone who will listen—including the press. Those who disagree with the organization can usually be identified at the first meeting; they are generally very outspoken about their feelings. It is best for the group to immediately incorporate these people into the core. Unfortunately, in most cases these people are not identified until much later. Despite this, still try to get them involved. In either case, give those people a job and a responsibility; make them feel important. Don't argue with them, but rather present their ideas, no matter how farfetched, to the decision-making body. If an idea is not viable it will be voted down, but the person presenting the idea will feel that someone has listened.

One of the best ways to guide people with a lot of energy is to make them street representatives, which gives them an important role and provides a productive focus for their enthusiasm. A street representative is in charge of a certain number of homes in the area; their duties could include calling those homes to remind the residents of upcoming meetings, circulating petitions, and distributing flyers. The leader of the organization should meet with the street representatives every two weeks to get feedback on the needs, concerns, and feelings of their area of the community. Everyone benefits. Responsibilities are shared so that the core group does not end up doing everything, and more people become directly involved.

Strategies

There are many options open to an organization that wants to apply pressure, educate the public, or relieve tension within a community. The leader, however, must be careful not to overuse or exhaust the membership. Each protest must be carefully calculated, asking such questions as: "What do we want to accomplish with this action?" "Who is our target?" "Should we take this action now—or wait for a better opportunity?" Plan your activism carefully for a time when you can accomplish something, even if it is just an endorsement of an elected official. Your members will feel good about the success and become more willing to take part in the next action. Spread your activities out, as people become tired quickly; too many activities will cause fewer people to turn out when you need them most.

Your organization should not use violent protests or actions. A peaceful demonstation is just as effective and will not alienate your outside support. A Mother's Day Rally, for example, would be a good opportunity to express your fears and gain publicity and support.

The media is a vital resource, and thus they should be notified a few days in advance of the planned action. But in order for the media to cover the event, it should be media worthy; this is why a theme is necessary. The media likes to cover events on holidays, when government and business is slow.

During your protests, large signs and other props that graphically explain your plight should be used. When speaking to the media at your event, you must keep in mind that television and radio have very little spare air time. Your statement should be said in no more than three *short* sentences; anything longer will be cut. Print media is much different: It gives you a better opportunity to explain the issues.

Be sure to keep your signs simple and your statements concise. Talk and write in everyday terms. For example, it is better to say, "We have a high rate of birth defects," than to say, "Our birth defect rate has been found to be statistically significant with a threefold increase when compared to a comparable neighborhood."

There are many techniques for calling attention to your organization or cause. Although your group can learn from others' successes, it is beneficial to be creative and develop your own tactics. Some of the actions that have been used across the country include:

Picketing. This form of protest is used by almost every organization. You should first check with local government to obtain a permit, if one is necessary. The advantage of picketing is that it can be done anywhere—in front of a city hall, around a dump site gate, at the steps of the Capitol, or even at the home of the president of an industry or an elected official. Let their neighbors find out who is really living next door. Have fact sheets available to give out to motorists or pedestrians who pass by.

Rally. This is a good forum for bringing people together, especially if you can find a band or a person who will attract a crowd. It is also a good way to entice the elected officials who support you to give a statement, thus adding credibility to your group.

Symbolic Coffin/Motorcades. This is another action that is widely used, and there are several ways to go about it. When carrying the coffin to the "burial site," for example, you can carry it in a parade. In Massachusetts, a coffin was carried for five miles in a procession ending with a twenty-one-gun salute at a proposed dump site. At Love Canal, there was a motorcade with cars driving slowly, lights on, and signs on the sides of the vehicles.

Sit-Ins or Sleep-Ins. This is a time-consuming but very effective protest. Take your sleeping bags to city hall or another building and say, "Since it is not safe in our homes, we will sleep here, where it is safe." Don't just talk about it—really spend the night. A sit-in is similar; you can take over an office and explain that, since your home is unsafe, you are seeking refuge in the building.

Die-In. This method is particularly effective with corporations. Hooker Chemical Corporation (responsible for Love Canal) held an open house that two hundred Love Canal people attended. All protestors wore T-shirts that read, "Love Canal—Another Product of Hooker Chemical Corporation." When one person blew a police whistle, we all fell right where we were, across tables, on the ground, everywhere. This action was unannounced and made a considerable impact.

Prayer Vigil. This can be done at a church near the site. This is very useful in incorporating many people who do not want to be involved in other forms of protest.

Talking Outhouse. This is a good educational tactic that was used in Minnesota. The outhouse, equipped with a microphone and speakers, was placed outside the capitol. When someone walked by, the outhouse would ask if that person knew what the government was doing in Hector, Minnesota. Out of curiosity, many people talked to the outhouse and were educated about the problem. It is another creative strategy. Did you ever see people in three-piece suits talking to an outhouse?

Fund Raising

Fund raising is a very important part of any organization. Without funds, your organization cannot function successfully.

There are many ways to raise funds both inside and outside of the community. One idea is to pass the hat at your meetings, or hold fifty-fifty drawings at each meeting. A fifty-fifty drawing means you split the money collected between the organization and the winner. In many cases, the person who wins gives the money back to the group.

Selling T-shirts is a good means of fund raising. Not only are you making money, but you are conveying your message visibly.

In most small communities, local businesses will give direct donations to the organization. You can make an ad book, selling ad space to these small businesses. The ad book can contain a chronology of events in the front and the ads in the back.

A local supermarket may be willing to give a side of beef, or an appliance store might give a TV, that your organization can raffle off. The raffles can raise a few thousand dollars for the group.

A fund raising dinner-dance is another idea. This is even more lucrative if you can find an entertainer who will draw a crowd.

Block yard sales will raise small amounts of money with little effort. They are also a good chance to talk with others in the neighborhood, helping you to continue educating and updating the community.

Many church organizations are willing to give funds to local community groups. Your organization should approach church leaders and ask what the procedures are for requesting support.

Door-to-door canvassing can sometimes be used to raise funds inside and outside of the affected community. Before beginning, you should check to see if a permit is necessary. This takes a little more work than some fundraising techniques. Use the same type of approach as mentioned earlier for going door-to-door in your own community.

Conclusion

The strategies and methods described in this chapter have been successful in organizing citizens at the community level to get action in addressing hazardous-waste issues. Many of these approaches have also been used by other public interest groups to stop unnecessary highway construction, obtain jobs, improve public transportation, challenge discriminatory lending practices by banks, and protect building on land of significant community value. The value of community organizing around any problem or issue should never be overlooked. It is the power of citizens to organize and directly assert their influence that is the ultimate defense policy for protecting our environment.

7 Lobbying for the Environment

Brock Evans

During the past twenty years, I have had the good fortune to work with concerned citizens, in the Pacific Northwest in particular, and throughout the country, in addressing environmental and conservation issues. In representing citizen groups, you quickly discover that the resources at your disposal differ markedly from those that are available to industry. The game is the same, and the rules are the same; but the principles, and the ways of acting to accomplish your goal, differ.

Every section of our country has its own unique political and environmental culture. There is no one formula for environmental action that can be uniformly applied; however, there are some basic principles that are the same no matter who we are, where we are from, or what issue concerns us.

There are some basic axioms of citizen action that lead to success. When these are not followed, we usually fail. I will describe several of these principles of how citizens can operate to protect and preserve our environment that have worked in the past, and that I believe will still work in the future.

Don't Be Afraid

One of the first things that citizens become aware of when they decide to tackle an environmental issue is that the other side has a lot of money, a lot of power, and usually a lot of political connections. For example, it is rare that any citizen group meeting in someone's living room has the resources to match the public-relations firms, the lawyers, the media access, and the political experience of most polluting industries. It is quite natural to feel that we can't prevail in the face of all of that.

But don't be afraid. Although the opposition may have considerable money and power, as environmental leaders we have power as well. We have the power of the vote; we have the power of the people. The things that we believe in, such as wildlife, wilderness, clean water, and clean air, are values that are deeply shared by the rest of the American people. Whether or not these people belong to environmental or conservation organizations, they are with us. Our job is to get to these people and let them know what the issues are.

We are fortunate to live in a society where the power of an aroused citizenry counts for something. At the most basic political level, that power means votes to our elected representatives who must make decisions about the places and the things we love. The greatness of our political system, and the thing that should encourage anyone who wants to protect the environment, is that if enough citizens are sufficiently concerned, elected officials will be responsive to them. Although it is natural to be impressed and somewhat fearful of the concentrated power of the opposition in environmental battles, it is encouraging to remember that we have won thousands of battles in the past few decades through citizen power and in the face of all the money, all the odds, all the lawyers, and all the political connections of the other side. So, as a first principle, I encourage any citizens who would take action to protect the environment not to be afraid. We, too, have power.

Don't Quit

A second axiom of citizen action is to be persistent, and never let go. Important battles to protect the environment are often long, and staying power is one of the keys to victory. The beginning is often discouraging because those in power may be suspicious of what you say at first. After all, who are you? Why should your opinion be better than the established opinion?

So, from the beginning, it is clear that there is much to do. It is going to require education, hard work, talking to many people, and becoming involved in the political process to attract supporters and establish credibility with public officials. But this can be done, as the thousands of successful environmental battles that have taken place in all parts of our country demonstrate. However, it will take time. For example, efforts to establish wilderness areas in the Northwest, Missouri, Florida, and New England took ten to fifteen years. It took nine years to pass the Alaskan Land Bill. It took eight years to pass the Wilderness Act, and about ten years to establish strip-mining laws.

While it may or may not take this long to prompt action at the local and state levels, be prepared for the long haul. Don't quit if you lose the first time. Come back a second, third, or fourth time. Come back until you are done and you have won. Outlast the opposition. That is what significant citizen action is all about; and the long, passionate, and inspiring history of the environmental movement with its scores of successes shows that we can do it.

Know Your Facts

Our democratic system provides mechanisms that allow every interest to have its say and be heard. In my experience as an advocate of environmental and conservation values and protective legislation, I have observed two important characteristics of our system. The first is that politicians are almost always willing to hear what everyone has to say before making their final decisions on an issue. Second, I have found that our opponents are honest people with sincere beliefs, and that they have valuable and important things to say as well. While it is usually in their economic self-interest to pollute or exploit a wilderness area, it should be recognized that this is an interest that is important to them, their employees, and others in their community or industry.

I have also observed that these people are usually wrong in their perception that what we ask for will necessarily interfere with their goals. That is why it is important to listen carefully to what they have to say. Then we must go out and ascertain the facts, find our own experts, and develop objective and balanced arguments. Listen to what they have to say, and answer every point with a counterpoint—that is the secret. What we should attempt to do is to demonstate, at least to the satisfaction of public officials who will make the decision, that what we propose will be beneficial to the environment but will not have a significantly negative effect on industry.

We win our issues in proportion to the fullness of our preparation, the extent to which we listen to what opponents say, and our ability to know the facts better than anyone else. Information is power in the game of saving the environment. We can never win on rhetoric alone.

Understand the Political Nature of Everything

Environmental leaders need to understand the practical dynamics of our political system. By this I do not mean partisan politics so much as the processes in our system whereby politicians or appointed administrators who are responsive to politicians make decisions about strengthening or weakening environmental regulations, permitting or stopping pollution, establishing or closing wilderness areas, and the like.

Because our system operates according to these processes, we cannot fail to participate in them. It is not enough only to gather our facts and be determined to persist. We must also be prepared to go to public hearings and testify. We must organize for hearings and convince many people to speak out. We must write letters, make phone calls, and meet with public officials whenever an important decision is being made. Our opponents are going to do this, so we will surely lose out if we do not. It is as simple as that.

In addition to becoming involved in these ways, we should also be prepared to participate in electoral politics if need be. For example, it may be that a politician who chairs a powerful committee is implacably opposed to our values and positions. We are never going to convince him or her with all the letters or phone calls in the world. The only solution in such a situation is to work to remove the person from office by defeating him or her in the next election. It doesn't matter which party we support or oppose; but it does matter what each elected official does or doesn't believe. Therefore, we must also locate good potential candidates, urge them to run, support them, and work hard in their campaigns. Then we will have champions for our causes where we need them.

Some Basic Principles of Effective Lobbying

Once we have observed these principles, there comes a time when we have to walk the halls of our state house, county seat, or the U.S. Congress to engage in direct lobbying. For those who have never done it before, lobbying is not as scary or as difficult as it may sound. Lobbying is simply the basic art of communicating your views to your elected or appointed officials. You are letting them know how you feel, explaining why, and urging them to support your position. To demystify this process, it helps if you also realize that this is what officials are for—to listen to the views of people as a part of making decisions. This is what they are paid for; this is why they ran for office; this is their responsibility.

People who lobby perform an important function in our political system. Because there are so many complex issues in our society, no one, no matter how talented, can possibly learn them all. Therefore, respected lobbyists can have great influence because politicians and appointed officials learn to trust them and rely on their information.

What are the most important principles of successful lobbying? In particular, I have found the following four grounds rules to be of greatest importance, whether one is lobbying for an environmental organization, an industry, or any other group.

Be Presentable. This may sound like a very elementary maxim, but never underestimate its importance. It is particularly important for citizen activists to understand that violating this principle can dilute all their good intentions. Assume, for example, that you have a life style, way of dress, and values that are pleasing and acceptable to you and your friends. But if your style differs too much from the norms of society, you simply will not get very far when you meet with public officials. Therefore, I recommend that if you want to be an effective lobbyist, you should be sensitive to your personal style as well as the substance of your issues. Be clean. Dress neatly. It is preferable to wear a jacket and tie if you are male, a dress or appropriate slacks if you are female. Dress so that your appearance cannot detract in any way from your message.

Be Absolutely Credible. A good lobbyist never lies or stretches the truth. They always answer as honestly as they can, given the information that is available to them. The reason for this is simple: you are essentially asking an official to trust you and your information. If elected officials do vote as you wish and later find that you have misled them, you can be assured that while you influenced their vote once, you will never be able to talk with them again. They will never forget, and they will never listen to you. So, when you lobby an individual, tell them exactly how you feel and why you believe in what you do. Show them your facts and figures to back up your position. If you have done your homework, you will find that you can often impress them.

Those whom you lobby may ask you questions such as, "Will you explain what your opponent's arguments are?" Do that honestly and fairly; you will only gain their respect by doing so. Of course, there is nothing wrong with your pointing out your answers to these arguments at the same time; but never lie or stretch things. This is how to become trusted; and, as a result, your views will carry more weight.

Give Information Briefly. It is important to appreciate the fact that elected officials are very busy people with a lot of things on their minds. As much as they might like you or what you believe in, they cannot really give you much time. Consequently, when you provide them with information, be brief and to the point. It may have taken you five years to learn what you are talking about, but don't take five hours explaining it. Take five to ten minutes. If they are still attentive when you are done, then you can go on and elucidate, but leave it to them to ask you. By no means wander around the map while they get bored—they will not want to meet with you again. To reinforce your points, you can bring some material to leave with them. They or their staff may very well read it.

Get Information. Providing information to public officials to persuade them to vote your way is only half of the equation of effective lobbying. The other half is obtaining information from them. This means very simply that once you have presented your information, you may ask them questions, too, such as: How do you feel about what I said? What are you hearing from the other side? Can you support us? How do you think you will vote? Does anything bother you about this issue?

The answers to these questions are most useful and important. If, for example, they say that they are inclined to support you but they are not receiving any mail supporting your position, then you know what to do. You go back and make sure that mail supporting your position comes in from their district. If they say they are concerned about the impact of jobs in their district, then you know what to do here as well. You find information about this from the best expert you can locate and try to show that there will be no negative economic impact, or, if there is one, that it will be slight and outweighed by the other benefits of your position. In other words, make use of the information they give you so that you can help them deal with their political concerns. It is hoped you will deal with them so well that they will have no political problem in supporting you. That is, finally, what it is all about.

There are many more nuances, twists, and turns to the fascinating art of lobbying, but these are the principles that we have applied most uniformly across the country in many and varying kinds of battles to protect the environment. These principles have been extremely successful for those who came before us, and there is no reason to think that they cannot be equally successful to those of us who will carry on the great tradition of the environmental movement.

8 Environmental Testimony and Public Hearings: Some Questionable Assumptions

Judy B. Rosener

Although I believe in the potential value of public hearings, my experience indicates that many assumptions about hearings are questionable. Public hearings provide one of the most common forums in which environmental leaders can influence government policy. Government agencies, whether they are at the local, county, state, or federal level, frequently use public hearings to determine citizen views. Therefore, at one time or another, almost every environmental leader is likely to feel the need to testify at such an event.

Attitudes regarding the value of public hearings vary considerably. At one extreme, a number of public officials and citizens feel quite cynical about the process. To these people, public hearings are a waste of time since government officials really don't carefully consider public testimony in their decision making. At the other extreme, public hearings are seen as an efficient way of determining public attitudes, gaining additional information from the public, and obtaining thoughtful evaluation of policy alternatives. The truth is that public hearings differ, and any hearing might represent either of these extremes or fall somewhere in between them in terms of its value. Because of this, those who choose to testify at a public hearing must begin with the belief that the hearing at least has the potential of being worthwhile.

Citizens and leaders of interest groups are guided by a number of popular assumptions about public testimony and public hearings. For example, it is frequently assumed that the decision-making process following a hearing is based on a rational evaluation of relevant information and, consequently, substance is more important than style in testifying at a hearing. Also, it is often assumed that there are policy choices that are right because they are more consistent with the so-called public interest. Are there really "right" policy choices or one public interest?

What is my experience and why do I feel it qualifies me to question basic assumptions about public hearings? For twenty-five years I have been a citizen activist fighting environmental-quality and resource-management battles in California, battles that have taken me to many public hearings.

For over ten years I have studied and taught courses in public policymaking with a focus, and I prepared a doctoral dissertation on evaluating the effectiveness of citizen participation in public hearings and other governmental settings. For eight years I was a state commissioner, regulating and managing resources in the California Coastal Zone and sitting as a decision maker in thousands of public hearings. So I have experienced public hearings as a citizen trying to influence policymakers, as a regulator watching citizens trying to influence me, and as an academic researching whether or not participation in public hearings makes any difference.

Although I am about to question the validity of basic assumptions made about public hearings, I remain convinced that the public hearing as a concept, as a place where the public can be heard, is a worthwhile activity. My own research of almost two thousand hearings of the California Coastal Commission (a state regulatory agency that determines land use in the Coastal Zone) shows clearly that citizens can influence decision makers in public hearings.[1] Others have likewise shown that there are variations of the traditional public hearing resulting in participation that satisfies the needs and desires of citizens and public officials.

Checkoway has shown in his analysis of existing public-hearing research that there are shortcomings that need to be addressed by those designing and participating in public hearings.[2] I would like to add a shortcoming to his list: the need to question the following assumptions:

1.   Decision making is "rational."
2.   Substance is more important than form.
3.   There are right and wrong answers to public policy questions.
4.   There is such a thing as *the* public interest.
5.   All decision makers should be treated equally.
6.   Public officials always follow staff recommendations.
7.   Consensus building is more productive than conflict.
8.   If there is an opportunity to participate, those interested will participate.

Let's look at these assumptions one by one to see how they relate to public hearings.

Decision Making Is Rational

I first became an environmentalist in California while participating with others to save a public estuary in California from being traded to a private corporation. We assumed that the city, county, and state officials responsible for making the estuary decision would utilize a rational decision-making process. By *rational* we meant that all relevant information would be gathered, analyzed, and evaluated against some agreed-upon standards or criteria. Based on this assumption, we presented our case in the required public hearings in a manner that was consistent with our image of the rational decision maker. We provided quantities of information, outlined what we saw as public costs and benefits, and even suggested our idea of what the decision criteria might be. We expected that the materials we provided would be carefully read by each public official involved in the decision, and that the arguments of various interest groups would be given equal attention. Were we wrong! But not for the reasons you might suspect.

We were wrong because all relevant information can never be obtained in a way that is satisfactory to all interests. Those with the most resources are always able to provide more information than those with fewer resources. We were wrong because it is impossible for public officials to read the mountains of paper provided them given the time available. We were wrong because there are no agreed-upon standards or criteria by which environmental quality or resource management can be measured. But mostly we were wrong to assume a "rational decision-making process" because public policymakers are human beings; they vote their values!

Put differently, public-policy decisions are value judgments, even though they are often clothed in numbers and studies to make them appear rational. I say this after having carefully observed my own behavior and that of the other public officials who served with me as regulators on the California Coastal Commission. The problem is not that public officials would not like to have a logical, orderly, objective, decision-making process, but rather that the time and resource constraints on most public bodies do not permit this. Because there are no agreed-upon, clearly defined standards, decision makers find themselves faced with the need to interpret vague mandates. This means they exercise discretionary authority. Because of information overload and conflicting information, public officials need to find a hook for their decisions. That hook is most often a decision maker's own value system. Why, then, does voting one's values seem irrational? It is, in my opinion, because we assume that those with the authority to make decisions on behalf of others will act differently from the rest of us; we hope they will. This assumption is wishful thinking.

At a public hearing, information is presented and decision makers have to make choices where there are not always simple answers. The lesson to be learned is that only if the word *rational* is viewed as making choices consistent with a given value system can we say public policymaking is rational. So, savvy public-hearing participants should prepare testimony or remarks couched in terms that are consistent with the values of those they wish to influence.

Substance Is More Important Than Form

Anyone who has participated in an environmental organization knows about disagreements between those who want to concentrate a group effort on the substance of a public hearing presentation and those who feel it is the style or form of the presentation that is most important. Substance and form are not mutually exclusive; both are important. However, there is a widely held view that the legitimacy of a presentation is related to its substance. Much of this is based on a notion that the public-hearing record is what is important—that it is the record that ultimately makes the difference in voting outcomes. This belief jibes with the previous assumption that there is a rational decision-making process in which the record is carefully read and analyzed by all the decision makers. How carefully public hearing testimony is read and how important that testimony becomes are dependent on the kind of hearing, the kind of decision body, the kind of decision makers, and the need of participants to establish legal standing. To assume that a record of what is said in public hearings is always read carefully by all the decision makers is a mistake. There is no guarantee that it will be read at all! Having detailed remarks or information on the record may be emotionally satisfying to participants, but it may not be politically productive.

Another way in which the substance-over-form assumption is manifested is in the context of who does the talking in a public hearing. Frequently it is the president of an organization, or a scientist, or some other titled or official representative of an interest group who makes the presentation, rather than a person who is better suited to selling an idea or raising an issue. I sat through many mumbled, boring, lackluster presentations that could have been much better delivered by someone who understood the importance of pauses, graphics, and brevity. The *shotgun* presentation, which is characterized by making a large number of points in the hope that one will hit home, is often used in place of the *bullet* presentation, which puts forth one or two points that are well aimed and carefully timed. While the analogy is not perfect, those using the shotgun approach tend to feel substance is what's important. Those using the *bullet* approach opt for the importance of form and focus.

The problem of information overload is also related to the substance-form debate. Not only is much of what is said in public hearings repetitive, it is also frequently irrelevant. Testimony that contains numerous points cannot be fully absorbed by the listener. Therefore, the participant who provides testimony that is brief and to the point, and makes it easy for the listener to absorb, has a decided advantage. Ironically, a participant who modulates his or her voice effectively, uses analogies and metaphors, and has a sense of the dramatic can often be more influential than the expert who may better understand the issue. The lesson to be learned is that both substance and form are important, but participants have to understand that the purpose of participation in a public hearing is to convince the listener of some point of view. Getting the listener's attention is the first step toward accomplishing that goal! Getting a listener's attention is often more related to form than substance in the beginning.

There Are Right and Wrong Answers to
Public-Policy Questions

It is easy to believe that environmental quality or resource management are issues that have so-called right and wrong answers. Certainly it is right to demand clean air. But how clean is clean? Certainly it is wrong to create an adverse environmental effect. But what constitutes an adverse environmental effect? Having to decide what is right and wrong implies, as does the term *rational*, that there are standards by which right and wrong can be determined. Unfortunately, this is not the case. When citizens enter a public hearing assuming there is a right and wrong decision to be made, they are in trouble. In looking for right answers, decision makers rely on a number of factors in addition to their own values. They analyze the information being presented to them. They consider precedents, staff recommendations, the opinions of their friends and colleagues, poll results, and political realities. It doesn't take long to understand why the making of public policy is so difficult—it is because there are so few right and wrong answers.

I remember how surprised I was to discover myself voting against a Coastal Commission staff recommendation that a landowner on a very small beach-front lot donate a piece of that lot for public access as a condition of receiving a permit. I had come to the job of Coastal Commissioner committed to maximizing beach access. I had assumed that I would always vote for public access because that was the "right" thing to do. In this case there already was public access to the beach a short distance away. The question of how much access was needed had to be weighed against asking a property owner to donate a piece of a small lot in return for the right to build a home. Most public-policy making involves similar dilemmas. What is right in one instance turns out to be wrong in another. If this were not the case, there would be no need for commissioners. Legislators would pass laws and law enforcement officers would enforce them. In order for most laws to be enacted, they have to be vague so that legislators of differing political persuasions can agree on their value. It is left up to administrative agencies to decide how the laws should be interpreted. The delegation of discretionary authority to commissioners and hearing officers necessarily means that right and wrong will be decided differently depending on the particular decision makers. The lesson here is that right and wrong in environmental decision making, as in all public-policy making, are highly individual assessments.

There Is Such a Thing as the Public Interest

All of us at some time or another have used the term *the public interest* to describe a position or action. When asked to define *the public interest*, all of us stumble, for there are many public interests. There is no *the* public interest, and unfortunately the many public interests are sometimes in conflict. It is precisely when they are in conflict that decision makers have a problem. Participants in public hearings need to realize the difficulty that such conflicts pose for the decision maker.

The issue of conflicting public interests is often masked in legislative mandates that characterize broad public policies. For example, the California Coastal Act calls for the protection of fragile ocean tidepools at the same time that it calls for maximizing public access to the beaches, both under the guise of protecting coastal resources. Is it possible to protect tidepools when beaches are crowded with adults and children anxious to take home a crab or some other tidepool inhabitant? Is *the* public interest, the protection of the tidepools or maximizing public access? Participants in public hearings need to be careful that they do not equate their own interests with *the* public interest, unless they are prepared to support their position. Using *the* public interest as justification for a particular action is risky; a better way would be to state that there is *a* public interest in some decision outcome.

All Decision Makers Should Be
Treated Equally

It is often assumed by participants in public hearings that a county board of supervisors, a city council, or a state commission confers equal power on its members. While it is true that each member has only one vote, it is not true that each member has the same amount of power. Power comes from a great many sources, and power conferred by an institutional tie is only one source. Implicit in the assumption that decision makers should be treated equally is the notion that decision makers have equal power, and that all start from a neutral position on issues or projects. Of course, they do not. Many times I have been called by environmentalists and asked the addresses of all twelve coastal commissioners so they could be lobbied on a project. When I asked why all twelve were being contacted, my question has always met with surprise. With limited resources, public hearing participants need to realize while they should communicate with decision makers who support their positions, they need not spend time trying to reinforce strongly held views. Resources should be expended trying to influence those who have not made up their minds, or those who, because of special power, can influence other decision makers.

Public Officials Always Follow
Staff Recommendations

A number of studies suggest that public officials closely follow staff recommendations. It is assumed to be true.[3] Thus it follows that citizens, if they influence staff members, could avoid going to public hearings. Since most staff members reflect the views of those they serve (if they didn't, they wouldn't last very long), it can be assumed that there is a similarity between what public officials want and what staffs recommend. However, that does not mean that decision makers always follow their staffs.

The argument that bureaucrats have too much power is related to the fact that public officials depend on their staffs to process large numbers of written and oral communications, which form the basis of public-policy decisions. And it is often difficult to know whether the staff leads the public officials or vice versa. Based on observations of my own behavior and that of my colleagues, I suspect that we follow our staff in a selective manner. To test the validity of my observations, I did a study of the relationships between staff recommendations and voting outcomes in almost two thousand public hearings held by the California Coastal Commission.[4] I looked at staff recommendations in terms of approvals and denials. At the same time, I looked at the presence or absence of participants in the public hearings. Each project had an individual public hearing at which time all commissioners listened to the testimony, received a staff recommendation, and, after public discussion, voted on the project. Participants identified at the outset whether they were there to support or oppose a project. I found that when the staff recommended approval, commissioners voted approval 93 percent of the time; but when the staff recommended denial, the commissioners voted to deny only 55 percent of the time. When viewed in the aggregate, the commission followed the staff 85 percent of the time, a figure similar to that stated in other studies. However, it is clear that following the staff only 55 percent of the time on recommended denials suggests that it would be a mistake to assume that participants need only to influence the staff.

To examine whether participation was a factor independent of the state recommendation, I looked at whether or not the presence or absence of participants was related to the way commissioners voted. When the staff recommended approval and there were no participants in opposition, commissioners voted approval 89 percent of the time. When there were participants in opposition, the approval rate dropped to 66 percent. When the staff recommended denial and there was no citizen opposition, commissioners voted denial 11 percent of the time. When the staff recommended denial and there was citizen opposition, the denial rate increased to 34 percent! Given an overall 18-percent denial rate of projects, the data indicate that citizens who participated in opposition to development projects were able to double the number of projects denied as a result of their participation, irrespective of the staff recommendations. It becomes clear, then, that while public hearings are often perceived as being perfunctory and that it is staff recommendations that are the driving force in explaining decisions, participation can play an important part in counteracting staff recommendations.

Consensus Building Is More Productive
Than Conflict

Environmentalists have been blamed over the last few years for slowing down the policymaking process and creating unnecessary conflict. In response to this charge, environmental-mediation centers, conflict-resolution seminars, and new kinds of consensus-building participation techniques are receiving a great deal of attention. I intuit that consensus building is more productive than conflict. But what do we mean by *productive*? Sitting through thousands of public hearings and having to base my vote on what I heard in public hearings taught me that conflict, while unpleasant and uncomfortable, often generates new ideas and illuminates old ones. Conflict often seems to sharpen the positions of combatants. While I agree that conflict can be messy and needs to be managed, and that conflict resolution is important, it should not be assumed that conflict in and of itself is unproductive. Conflict attracts media attention, which serves to educate a wider audience than that which might attend a public hearing. That is something quiet consensus building seldom does. Both the opportunity to square off and the need to develop consensus should be goals of those designing and participating in public hearings.

If There Is an Opportunity to Participate,
Those Interested Will Participate

The assumption that citizens will participate if opportunities to participate are provided is linked to traditional political theory. That is probably why public hearings are so often mandated; they symbolize the democratic process. But an opportunity without resources to take advantage of that opportunity is a hollow promise. As a commissioner, I quickly learned that participation in public hearings is a cost of doing business to those with a financial interest in decision outcomes. I learned that delaying votes and discussions is a ploy sometimes used by those who are aware that citizen groups can be worn down by having to expend funds for traveling to and attending hearings. I learned that the cost of such items as obtaining documents, parking, baby sitters, transportation, and phone calls makes participating difficult for many individuals and groups who, while affected by decisions, have no ability to write off their participation costs or pass them on to others. Certainly the problem of having to pay for experts on technical issues is a cost that few citizen groups can afford.

On the federal level under the Carter administration, there was an effort to provide what was termed intervenor financing, which would help defray costs associated with providing experts in complex issues, such as the regulation of nuclear plants.[5] These efforts have been discontinued under the Reagan administration, as have attempts to experiment with innovative kinds of public hearings. This is unfortunate, since the past demonstrates that attempts made to lower participation costs result in more people being able to take part in public hearings.

In addition to the costs of participation, knowledge of participation opportunities is often limited to particular groups or individuals, leaving affected—but unidentified—people unaware of opportunities. There is a need to pay special attention to making sure that opportunities to participate in public hearings are widely advertised in a timely manner, and that the cost of participation is kept to a minimum. Providing participation opportunities in public hearings is not enough.

What Can Be Learned from Questioning Assumptions about Behavior in Public Hearings?

What we learn from questioning basic assumptions about behavior in public hearings is that things are not always as they seem; that it is important to think about the specific setting of a given public hearing prior to making plans about how to participate in it. Public hearings differ in structure (formal or informal), in kinds of interaction (one-way or two-way), in types of listeners (decision-making bodies), in costs of participation (proximity to participants, information needs, time constraints), and in the complexity of the issues (local planning decisions or nuclear-power-plant siting).

In order to be effective, those involved in public hearings need to pay attention to how these differences affect opportunities to participate, and the type of participation that will be the most effective. They need to think about the decision makers, and how best to influence them. Each public hearing is different. Participants need to pay attention to the fact that public officials are like the rest of us: they are driven by a set of values; they are subject to information overload; they have no absolute standard by which to determine right and wrong; they are not equally powerful; they may or may not follow the advice of their staff. A good way to predict how a decision maker will behave at a public hearing is to put yourself in his or her place. How would you react to your own presentation at the end of eight hours of listening to others? How would you sort out information? How would you receive ideas that conflict with your own values? How would you determine *the* public interest? How would you react to participants who disagree with your staff?

The public hearing should be viewed as a challenge, not a chore. It's a game that anyone can play. Those who understand the purpose of the game and its rules, and those who take the time to understand the motives and values of the players, tend to win. A local newspaper reporter once wrote an article about the California Coastal Commission in which she stated that after observing the commission for some time she noticed that those participants who were very young or very old, those who were handicapped or pregnant, and those who had accents or appeared right before a meal break or at the end of a meeting seemed to be the most successful. I don't agree with her entirely, but she was right about the fact that decision making in public hearings is not predictable or objective. Participants who give thought to the total public-hearing experience, not merely the information they wish to convey, are the ones who appear to have impact.

Notes

1. Judy B. Rosener, "Citizen Participation in an Administrative State: Does the Public Hearing Work?" (Ph.D. diss., Claremont [California] Graduate School, 1979).

2. Barry Checkoway, "The Politics of Public Heaings," *Journal of Applied Behavioral Science* (1980), 4th ed., pp. 566–82.

3. Irving Schiffman, "The Limits of Local Planning Commissions," *Institute of Government Affairs* (Davis, California: University of California, 1975).

4. Judy B. Rosener, "Citizen Participation in an Administrative State: Does the Public Hearing Work?" (Ph.D. diss., Claremont [California] Graduate School, 1979).

5. Joan Aron, "Citizen Participation at Government Expense," *Public Administrative Review* (September/October, 1979).

Environmental Mediation: Advantages and Disadvantages

Douglas James Amy

Environmental disputes have become a permanent feature of the U.S. political landscape. Given the inherent differences between environmentalists and those dedicated to promoting continued growth in industrial and resource development, there seems to be little chance that these controversies will go away. Unfortunately, it seems that our political system has yet to devise an efficient way of resolving these disputes. Congressional policymaking often seems only able to produce vague and ambiguous environmental laws that merely temporarily dampen environmental conflicts. Typically, these disputes quickly resurface during the implementation phase of these policies. Attempts to resolve the issues at the administrative level often prove inadequate as well. Administrative policy decisions are perceived as being arbitrary—public hearings, ostensibly designed to facilitate public participation in the resolution of these disputes, are often seen as charades. Administrative decisions seem inevitably to alienate one side or another, and thus environmental controversies usually end up in court. But litigation as a way of resolving environmental controversies has also met much criticism. Court battles are very expensive and often result only in extensive delays and legal standoffs. In addition, court decisions tend to be narrow and piecemeal—a very poor way of trying to create a coherent set of environmental policies.

As a response to the growing dissatisfaction with these traditional approaches to resolving environmental controversies, there has emerged in recent years a new, less confrontational approach to resolving these disputes—a process known as environmental mediation. In general, the term *environmental mediation* is used to refer to the use of neutral mediators to facilitate the negotiated resolution of environmental controversies. But in actuality, there are a number of somewhat different dispute resolution techniques that fall under the heading of environmental mediation.[1] One approach, usually called joint problem solving by its practitioners, addresses itself to resolving differences between *potential* adversaries before an actual conflict develops over an environmental issue. Another approach, known as policy dialogue, brings together national leaders in the field of the environment and business to discuss wide-ranging, national environmental policy problems. The National Coal Policy Project conducted at Georgetown University during the 1970s is an example of this approach.[2] But by far the most common (and successful) approach to environmental mediation is one that addresses itself to environmental disputes that are site specific and that have been going on for some time. Typically, representatives from the local organizations involved in the particular dispute (environmental, business, and governmental) sit down together with a neutral mediator, and over a period of months negotiate a settlement to the dispute. Unlike an arbitrator, the mediator has no power to impose a settlement—all settlements must be voluntarily agreed upon by all the parties involved in the particular dispute. The principle role of the mediator is to use his or her skills to diffuse hostility, clear up miscommunications, and integrate the demands of the various parties into a compromise solution.

The Growth of Environmental Mediation

The earliest experiments with environmental mediation began in the mid 1970s.[3] Credit for first bringing the tools of mediation to bear on environmental issues usually goes to Gerald Cormick and Jane McCarthy. While at the Community Crisis Intervention Center in St. Louis in 1973, Cormick and McCarthy had a series of exploratory conversations with environmentalists, lawyers, industry representatives, and government officials; their dialogues revealed industry's increasing frustration with the impasses caused by confrontational approaches to environmental disputes, and a strong interest in the notion of extending mediation techniques into this area. After securing grants from The Ford and Rockefeller Foundations, Cormick and McCarthy began to examine more seriously the possibility of mediating environmental disputes and eventually became involved in mediating a dispute in Washington state between environmentalists, farmers, developers, and public officials over the damming of the flood-prone Snoqualmie River.[4]

After the successful completion of this initial mediation project, Cormick moved to Seattle and formed the Office of Environmental Mediation (now the Institute of Environmental Mediation) and took on several more mediation efforts in the Northwest. Cormick's work soon garnered national attention, and interest in this approach began to grow. Today, environmental mediation has moved out of the experimental stage and is rapidly becoming institutionalized and professionalized. There is an increasing number of small mediation institutes around the country involved in training mediators and offering mediation services. Besides Cormick's office, there is also the Institute for Environmental Negotiation at the University of Virginia, Charlottesville; The New England Environmental Mediation Center, in Boston; the Center for Environmental Problem Solving, in Boulder; and others.[5]

There are also several books on the subject of environmental mediation, describing both techniques and case studies. Two of the most prominent are Laura Lake's edited volume, *Environmental Mediation: The Search for Consensus* and *Mediation of Environmental Disputes*, by Scott Mernitz. Another sign of the coming of age of mediation is the emergence of a quarterly newsletter, *Resolve*, aimed at those who are involved in environmental dispute resolution and put out by the Conservation Foundation in Washington, D.C. The Conservation Foundation has also been active in sponsoring conferences for practitioners of environmental mediation and others interested in these techniques.

Government Interest in Environmental Mediation

During the last several years, both federal and state governments have shown increased interest in using mediation to settle their differences with environmental groups. In California, for example, a Palo Alto firm, FORUM, has been engaged to conduct a two-year experiment on how to use conflict-resolution techniques in fashioning recreational-resource-development policy for the state park system. On the federal level, one manifestation of interest in mediation has been the participation of federal officials in a series of workshops put on by such advocates of mediation as Clark-McGlennon Associates, of Boston, and the American Arbitration Association. In these workshops, personnel from the National Park Service, the Federal Highway Administration, the U.S. Geological Survey, the Council on Environmental Quality, and other agencies have been made more familiar with the various techniques of mediation and their benefits.[6] Several of these agencies have also been able to use mediation to fashion successful compromises with environmental groups. For instance, the Forest Service used mediation to reach a compromise with environmentalists in a conflict over timbering and the protection of chinook salmon in Idaho's Salmon River. Also, in Wisconsin, the Army Corp of Engineers, the EPA, and the Fish and Wildlife Service all participated in fashioning a mediated settlement to a conflict over restoration of a wetland.[7]

One case in particular, involving the Forest Service and environmentalists in South Carolina, serves to illustrate how environmental mediation can help to generate mutually acceptable solutions to what could become very long and costly disputes. In the Francis Marion National Forest in South Carolina, there is a marshy area of several thousand acres known as Ion Swamp. The swamp is one of the last known nesting places of the Bachman's warbler—described in *Peterson's Field Guide* as "the rarest North American songbird." Environmentalists in the area became upset when, in 1976, the U.S. Forest Service announced its intention to permit logging in the area. At first it appeared that the ensuing dispute between the environmentalists and the Forest Service would go the same route of many such controversies and become entangled in a long and expensive court battle. However, Robert Golten, an attorney for the National Wildlife Federation, investigated the situation and suggested that the parties try to mediate their dispute. A mediation panel was chosen and the parties agreed to a moratorium on both litigation and lumbering during the mediation process. After several months of dialogue, field trips, and research, the panel recommended a compromise: that lumbering be permitted, but not in the bottom land areas that served as nesting places for the warbler. Both the Forest Service and the environmentalists quickly accepted this proposal.[8]

The Advantages of Environmental Mediation

The Bachman's-warbler case illustrates a number of the advantages that environmental mediation has over other approaches to dispute resolution. The advantage most often mentioned by advocates is that mediation allows the parties to avoid long and costly court battles. Litigation can often be prohibitively expensive—and this can be especially troublesome for local environmental and public-interest groups. In addition, valuable time can be lost in litigation because of court backlogs and time-consuming legal procedures. Mediation projects typically take much less time than court cases, though the time required may vary considerably from one dispute to another. There are occasions, of course, when delay may be a desirable tactic, but often delay means only continued environmental damage, and in those circumstances the speed of mediation makes it an attractive option.

Litigation can also be a very risky venture. Typically, court decisions find in the favor of one party or the other—so there is the real possibility that one could lose everything. In contrast, mediation offers the opportunity for all parties to achieve at least some of their main objectives. It assumes that not all environmental controversies are black and white, and that negotiations can reveal a middle path that will satisfy all parties. Unlike litigation and administrative decision making, this approach attempts not just to *decide* an issue, but to *resolve* it by finding a compromise that everyone can live with. This is an important advantage. Administrative or court decisions often only temporarily settle a question. Because the losing side of the dispute may come away with little or nothing from such decisions, they are inclined to continue pursuing their claims. Consequently, even if environmentalists win their case, the basic dispute often continues and they may easily find themselves back in court. In contrast, successful mediation agreements usually contain something for each party involved, and this lessens the possibility that the dispute will resurface.

Litigation often fails to resolve disputes, not only because of the all-or-nothing nature of its decisions but also because the grounds for decision frequently do not address the issues that lie at the heart of the controversy. In court proceedings, the only issues that may be addressed are those that are deemed litigable, and what is litigable may differ greatly from what is really at issue in the case. For example, in the case of the Alaskan pipeline, many court battles centered around narrow right-of-way issues, while the real dispute was clearly over whether the pipeline should be built at all. Similarly, because many environmental cases are based on the Administrative Procedures Act or the National Environmental Protection Act, they often revolve around procedural issues—how a decision was made, rather than its substance. When these cases are decided on procedural grounds, the substantive issues remain unresolved and again there is the likelihood that the dispute will resurface. However, since environmental mediation is an informal process, the participants are free to agree on what issues need to be addressed, and this presents the possibility that the central issues separating the parties may be finally resolved.

The informal nature of the dialogue that takes place in mediation has other advantages as well. Prolonged exchanges on a variety of substantive environmental issues can serve as an opportunity for environmentalists to environment-educate their opponents. Environmental disputes often are based, at least in part, on faulty information and mistaken assumptions—and in many cases, the parties in a dispute have not been forced to question or defend their own information and assumptions. However, the give-and-take dialogue in mediation often serves to expose these assumptions and submit them to critical questioning. As a result, many of the parties in mediation efforts, including environmentalists, have reported that during negotiations they became aware of previously unnoticed weaknesses and inaccuracies in their own cases. When this occurs, mediation can become more than just splitting the difference between two dogmatically held positions; it can become a learning process where the parties actually modify parts of their original positions. While this process of education certainly does not take place in every mediation effort, it is an advantage to the process of mediation that should not be underestimated.

Finally, mediation may also serve as an opportunity to improve the image of environmentalists with their adversaries and with the public at large. For example, participants in prolonged mediation efforts often report coming away with a new respect for their opponents. In one negotiation project involving coal mining, coal executives were at first dismayed to learn that one of the environmental representatives was to be a university geologist who was considered a long-time enemy of the coal industry. But after a number of encounters, the executives came away greatly impressed with the geologist's expertise and even-handed attitude, and one even expressed a willingness to consult with him in the future.[9]

Taking part in mediation may also improve the image that environmental organizations have with the public. Continual pursuit of confrontational tactics has often contributed to the portrait of environmentalists as extremists or intransigent naysayers. A willingness to negotiate, on the other hand, helps show that environmentalists can be flexible and reasonable in their demands. Of course this public-relations advantage is not in itself a sufficient reason to choose to mediate; and there are certainly times when being intransigent and unreasonable is the only reasonable path to take. But it is worth keeping in mind that mediation can serve to augment the credibility of environmentalists with the public and with policymakers, and that this can be an advantage in future attempts to create coalitions and participate in policymaking.[10]

Political Pitfalls in Environmental Mediation

Despite the advantages that mediation sometimes offers, there have been persistent suspicions about this process in the environmental community. Some have charged that the primary purpose of mediation is to distract or coopt environmental groups. Others have questioned whether it is appropriate for environmentalists to be "cooperating" with the developmental interests and have argued that compromise agreements may only serve to help legitimize environmentally destructive projects. These suspicions have been fueled by several stories in environmental publications that have pointed out that "industry clearly likes the idea" of mediation, and that some mediators' efforts and some mediation centers have been heavily funded by corporate foundations like the Ford and Rockefeller Foundations, and corporations with poor environmental records, like Dow Chemical and Union Carbide.[11]

But are these suspicions justified? Clearly, environmental mediation is not always cooptive. As the case of Bachman's warbler illustrates, mediation can at times produce agreements that satisfy the interests of environmentalists. Nevertheless, it is equally clear that the possibility of cooptation is widespread in mediation, and that environmentalists would do well to be aware of the political pitfalls inherent in this process. There are many of these pitfalls, but three warrant special attention: (1) the possibility of being "seduced" by the mediation process, (2) the possibility of wasting one's time negotiating with an opponent that is not bargaining in good faith, and (3) the possibility that a compromise solution is not appropriate in a given dispute.[12]

The possibility of seduction is rooted in the fact that opposing negotiators have a tendency to become friendly with each other over the course of a long mediation effort. Mediators often work hard to encourage participants to overcome their initial hostilities and the negative stereotypes they may have of each other. Also, prolonged negotiations often require spending long hours with one's opponents, and some mediators even encourage negotiating teams to share meals with each other. The natural tendency in such situations is for a kind of personal bond to develop between the people involved in the negotiations. Some have even reported developing feelings of friendship and trust for their opponents. Of course, this does not always occur; but the danger is that such feelings could get in the way of effective bargaining. And indeed, there have been several instances where an environmental group has accused its representatives of being too soft on the opposition and compromising too easily.

This kind of problem is most likely to take place when the negotiating team has no real experience in hard-nosed bargaining—as is often the case with local environmental groups. In addition, some observers, like James Benson, of the Institute for Ecological Policies, have suggested that another part of this problem may lie in the fact that environmentalists are simply too nice!

> Environmentalists are not really very feisty people, and when they get around a bunch of high-power corporate types they don't want to be unreasonable. They don't want to get into arguments. They want things to be peaceful and gentlemanly. . . . they get carried away and really may lose sight of what it is they may be trading off. They may be trading off something in the negotiations that others may be unwilling to trade off.[13]

Another typical pitfall in environmental mediation is becoming involved with an opponent who is not bargaining in good faith—a situation in which one's adversaries are negotiating primarily to gain concessions but are unwilling to make significant concessions themselves. This is a perennial problem in all forms of negotiations, and it is notoriously difficult to distinguish between tough bargaining and bad-faith bargaining. In some cases the lack of good faith is clear, as when an opponent constantly stalls for time and offers no counterproposals.[14] But other cases are much more ambiguous, as when an opponent is enthusiastic to negotiate, but only wants to discuss the peripheral issues, not the basic ones. This is a tactic that has often been used, for example, by developers of shopping malls. Frustrated in the past by local-citizen opposition to these massive shopping centers, some developers have now begun actively to seek out negotiations with local neighborhood and environmental groups that they believe might object to their project. But while these developers are willing to negotiate such issues as lighting, access roads, building height, and landscaping, they are usually unwilling to put the basic issue of whether the mall itself is desirable on the bargaining table. As one environmental lawyer has complained:

I was involved in a case in which a company wanted to build a massive regional shopping center in a suburban county. It would have effectively de-
stroyed what had been accomplished downtown—a classic case. Of course
we continually wanted to negotiate. But the developers wanted to negotiate
things like how to make it pretty and how many parks they would give. . . .
I refused to negotiate until they agreed to negotiate on the basic land use
question. . . .but they were unwilling to put that issue on the table.[15]

Unfortunately, this less than forthright approach to mediation has also
been encouraged by those who have been advising federal officials on how
they can use environmental mediation. One example of this is found in a recent government report entitled *New Tools for Resolving Environmental
Disputes: Introducing Federal Agencies to Environmental Mediation and
Related Techniques*—a document prepared for the Council on Environmental Quality (CEQ) by Wendy Emrich and Peter Clark. In this report,
the authors attempt to soothe the fears of government officials who suspect
that mediation means that they will have to negotiate over basic agency
policy and thus give up some of their policymaking power. Emrich and
Clark assure officials that they need not "fear the loss of agency authority
in the context of group negotiations," that "a properly managed mediation
guarantees retention of agency authority; it does not challenge or weaken
it."[16] A properly managed mediation turns out to be one in which the agency
determines which issues are up for negotiation and which are not.

In the words of the report, "at the outset, the mediator requires the
agency to outline its specific constraints—regulatory, political, economic—
under which it must operate and within which any final agreements must
fall."[17] By using this notion of constraints, an agency is able to narrow the
range of policy options available for negotiation. For example, at the beginning of negotiations, an agency may cite its budgetary constraints to
eliminate from the negotiations consideration of policy alternatives that are
more costly than its current plan. Or an agency might stipulate that it has a
legislative mandate to follow a certain policy—that its hands are tied in that
regard—but that it is more than willing to negotiate *how* those plans are to
be implemented. By using the notions of constraints in this way, an agency
would be able to predetermine the scope of negotiations and thus ensure
that only relatively minor issues are addressed in the mediation effort. This
approach allows officials to give the appearance of being flexible and willing to compromise, while in fact leaving many of the most basic issues un-
addressed.

It is revealing that some advocates of environmental mediation (including
the authors of the CEQ report) use the phrase *conflict-management technique*
to describe environmental mediation. This terminology should alert environmentalists that offers to mediate a dispute are sometimes best seen as attempts to "manage" citizen groups that could prove to be disruptive to the

plans of public and private developers. This intention to use mediation to manipulate rather than improve the quality of citizen participation in environmental policymaking usually remains carefully concealed; but there are times when it is inadvertently made more explicit, as in the CEQ report mentioned above, where the authors point out that "approaches to conflict management . . . need not be confused with more requirements for public participation. They key word is *management*. In cases where public groups are fighting with the federal government, better conflict management means better control of the participation process."[18] Thus, as this statement illustrates, mediation (like public hearings) may sometimes be used by government agencies to give the illusion of participation, while in fact functioning primarily as a forum for officials to push for their own policies.

The third major pitfall involved in environmental mediation is the possibility of an environmental group being lured into mediation efforts where compromise solutions are in fact unlikely or even undesirable. For example, some mediation advocates have suggested that environmental groups could benefit from participating in mediation efforts designed to resolve disputes over nuclear reactors. But as a rule, mediation efforts are usually unproductive in situations where policy decisions are of an all-or-nothing nature. In the case of a nuclear reactor, one either builds one or one does not—and the possibility of finding a true middle ground is quite small. When disputes have this all-or-nothing character, environmental groups would do best to be quite suspicious of offers to mediate. In such situations, mediation is most likely being employed as a tactic to distract environmental groups or to delay their taking other action.

There are other conflict situations in which a compromise may be feasible but undesirable. As James Crowfoot points out, there are times when negotiated compromises only serve to legitimize projects that are inherently environmentally destructive.[19] This is a constant possibility in environmental mediation, because the typical compromise between environmentalists and developers is agreement on some form of "responsible development"—continued development with environmental safeguards. In some cases, however, the central issue is not whether a development plan is responsible or not, but whether development should take place at all. Most environmentalists, for instance, would oppose oil exploration in national parks and wilderness areas irrespective of how carefully it is carried out. There are, then, conflicts that involve matters of principle in which compromise—and thus mediation—may be inappropriate. These principles may be moral, environmental, or even legal, as when one does not wish to compromise over the right of due process in environmental decisions. In these situations, as in situations where the decision is of an all-or-nothing nature, engaging in mediation may be largely a waste of time; or even worse, it may inadvertently result in the compromising of dearly held principles.

Some Guiding Principles

The presence of political dangers in environmental mediation should not imply that this is an option that should be avoided. To some extent, these problems can be mitigated by engaging a competent professional mediator to run the negotiatons. Politically sensitive mediators, like Gerald Cormick, are well aware of the political pitfalls involved in mediation and may, for example, recommend against pursuing mediation in cases (as with nuclear power plants) in which compromise is very unlikely. But while mediators can be very useful in assessing the potential for useful negotiations, it is important that an organization not rely on them too heavily in determining whether or not to mediate. That is a decision that must be made by the organization itself, after careful analysis of all of the potential advantages and disadvantages that mediation may offer in their *particular* situation. In reaching that decision, the following set of questions may serve as a set of guiding principles. They identify the minimum set of issues that an organization would want to discuss before deciding if they are ready and able to pursue a mediated solution to their dispute:

1. Have you exhausted your alternatives? While mediation may be attractive, it should not be your first option. It should be your last. If there are other approaches, such as litigation, that promise a good chance of victory, are within the resources of your organization, and seem likely to lay an issue to rest, they should be pursued first.
2. Do you have experienced negotiators to represent you? To effectively bargain and avoid cooptation, an organization must have negotiators familiar with the experience and tactics of bargaining. If that experience is not readily available, are you willing to take the time and expense necessary to acquire it?
3. Is this particular dispute one on which the organization should compromise? Is a compromise solution feasible? Is it desirable? (Also, do you have a choice? Compromise may turn out to be very attractive if the only other option is probable defeat.) It is critical that an organization generate substantial agreement on which issues are negotiable and which are not. Pursuing mediation without such agreement runs the risk of creating a serious split in the organization.
4. Are the members of the organization willing to put large amounts of time, work, and money into the mediation effort? Mediation is usually much cheaper than litigation, but it can still tax the resources of a small organization. Therefore, there must be a strong commitment by the membership. Otherwise you may be bargaining at a disadvantage. If, for instance, the organization is unwilling or unable to hire the necessary technical expertise to pursue its case, it will inevitably be outgunned

at the bargaining table by opponents who have better access to technical information and scientific expertise.

5. Are the negotiators for your opponents' organizations empowered to make policy decisions? Are the opposing companies or government agencies committed to abide by the agreements fashioned by its representatives? This is not always the case, and there have been a number of instances where environmentalists have spent months in thrashing out an agreement with lower-echelon officials only to have the agreement later repudiated by higher officials.

6. Are all interests affected by an issue being included in the negotiations? Mediators sometimes like to keep the number of parties involved as small as possible. But if there are other environmental or citizen groups with a different interest or position on the dispute, you should insist that they be included in the negotiations. Without full participation of all interested groups, it is unlikely that the final agreement will truly embody the public interest. Also, your group will then not have to shoulder the added burden of having to try to speak for these other groups.

7. Is there evidence that your opponents are willing to negotiate in good faith over the basic issues involved? If they make frivolous demands, engage in contant stalling, refuse to make counterproposals, or exhibit a "take it or leave it" attitude, these are all signs that mediation may be a waste of time. However, the initial insincerity of your adversaries should not necessarily preempt the mediation option—for it merely brings us to the final point to consider.

8. Are there ways that you can increase the likelihood that your adversaries will negotiate in good faith? In other words, can you create a political situation in which mediation becomes the most attractive option for your opponent? In practice, this usually depends on whether your organization can demonstrate the ability to interfere with or delay a company's or agency's goals. Unless you have that power, your adversaries will have little reason to negotiate seriously. As Cormick points out, one of the essential prerequisites of a successful mediation effort is the ability of the parties involved to frustrate each other. "The parties in a dispute will be willing to enter with 'good faith' into negotiation-mediation to the extent that they are unable to act unilaterally in what they perceive to be their own best interest."[20] For environmentalists, one of the most common ways to frustrate the unilateral actions of agencies has been through litigation; and in fact most successful mediation efforts have tended to take place after litigation has been initiated. The specter of a long and costly court battle has a way of motivating the parties to bargain and compromise. As a general rule of thumb, then, environmentalists should be skeptical of offers to mediate

that come early on in a dispute, before litigation or delays have begun. When an opponent is anxious to mediate before the dispute has come to a head, it is often an indication that cooptation rather than serious negotiation is on the agenda.

In the end, those environmental groups that will do best in mediation are those that see mediation not as a way of avoiding confrontation and conflict but as a final stage in a long process of political struggle. It must be kept in mind that mediation is a *political* process, and that like all forms of politics, the key is power. This means that the first priority of environmental groups should not be to seek out negotiations, but to mobilize and demonstrate their political power—whether it be through litigation, lobbying, demonstrations, or other activities. Only those environmental groups that can negotiate from a position of strength will be able to resist cooptation and use environmental mediation most effectively.

Notes

1. For more lengthy descriptions of the various forms of environmental-disputes resolution, see Howard Bellman, et al., "Environmental Conflict Resolution: Practitioners' Perspective of an Emerging Field," *Environmental Consensus* (Winter 1981), pp. 1–7.

2. For a description of this project, see Tom Alexander, "A Promising Try at Environmental Detente for Coal," *Fortune*, (Feb. 13, 1978).

3. For a short history of environmental mediation, see Scott Mernitz, *Mediation of Environmental Disputes* (N.Y.: Praeger Publishers, 1980), esp. chapter 5.

4. For an account of this project, see Gerald Cormick, "Mediating Environmental Controversies: Perspectives and First Experience," *Earth Law Journal* (1976), 2:215.

5. For a list of mediation institutes and current mediation projects, see current issues of *Resolve*, available free from the Conservation Foundation, in Washington, D.C.

6. *Workshop on Mediation* (Washington, D.C.: Resource and Land Investigation Program, Geological Survey, U.S. Department of the Interior, 1980).

7. For a description of these and other cases, see *New Approaches to Managing Environmental Conflict: How Can Government Use Them* (Washington, D.C.: Council on Environmental Quality, 1980).

8. Ibid.

9. Alexander, p. 99.

10. For more on this point, see James E. Crowfoot, "Negotiations: An Effective Tool for Citizen Organizations?" (Helena, Mont.: Northern

Rockies Action Group, 1980). This is one of the most politically insightful essays on mediation.

11. Geoffrey O'Gara, "Should the Marriage Be Saved?", *Environmental Action* (March 11, 1978), pp. 12–14.

12. For a more lengthy analysis of these potential problems, see Douglas James Amy, "The Politics of Environmental Mediation," *Ecology Law Quarterly*, vol. II, no. 1. pp. 1–19.

13. Telephone interview with James Benson, November 17, 1981.

14. For more on this point, see Crowfoot.

15. Northern Rockies Action Group, *Selected Transcripts from the NRAG Conference on Negotiations,* NRAG Papers (Fall 1980), p. 13.

16. Wendy Emrich and Peter Clark, *New Tools for Resolving Environmental Disputes: Introducing Federal Agencies to Environmental Mediation and Related Tehniques* (Washington, D.C.: Council on Environmental Quality, 1980), p. 12.

17. Ibid., p. 12.

18. Ibid., p. i.

Further Reading

Clark, Peter and Wendy Emrich. *New Tools for Resolving Environmental Disputes*. Washington, D.C.: Council on Environmental Quality, 1980.

Cormick, Gerald. "Intervention and Self-determination in Environmental Disputes." *Resolve* (Winter 1981).

Dowie, Mark. "Atomic Psyche Out: The Nuclear Industry's Strategy to Divide and Destroy the Opposition." *Mother Jones* (May 1981).

Environmental Mediation: An Effective Alternative? Washington, D.C.: The Conservation Foundation, 1978.

Lake, Laura, ed. *Environmental Mediation: The Search for Consensus*. Boulder, Colo.: Westview Press, 1981.

Susskind, Lawrence and Alan Weinstein. "Towards a Theory of Environmental Dispute Resolution." *Environmental Affairs* (1980), 9:311.

10 Mastering the Media

Nancy Wilson Anderson

If there is one practical political maxim that every environmental leader should remember, it is that the media is the greatest tool in educating the public and advocating environmental protection. Few environmental battles have been or can be won without coverage by print or electronic media or both. Whether or not environmentalists feel comfortable working with the media, an environmental advocate cannot be effective without knowing how to use it. Further, environmental activists cannot afford to assume that the media is responsible for going out and digging up stories about environmental issues.

Keeping the media up to date on environmental issues is difficult. We cannot assume that all reporters have the background necessary to understand the technical information presented at a public hearing on, for example, the potential health effects of a local hazardous-waste site. They are not all cognizant of the importance of tropical rain forests, wetlands, and flood plains, or the effect of acid rain on fish habitats and forests. For this reason, spokespeople for trade associations such as the National Salt Institute, the American Petroleum Institute, and the bottling industry can have a field day making use of the media to get their message before the public, without the public receiving an explanation of the other side of the issue—unless environmentalists speak up and ask for equal time.

The media supplements such educational tools as courses, workshops, conferences, and demonstrations. Almost everyone reads a newspaper, watches TV, or listens to a radio. The role of the media is, as Mark Twain once said, "to carry light to all corners of the globe. . . ." In order to do this, they must give comprehensive coverage to what is happening in their area and all over the world. While they receive many more announcements and news items than they can possibly have space or time to use, any informed, well-organized, and active citizen group can command their attention.

Background Information

One of the best principles to guide your work with the media is to know that you need the media and the media needs you. Their job is to provide information and news to readers, listeners, and viewers, and this must be done every day—sometimes four and five times a day. They cannot simply announce that "there is no news today," and have the radio go silent, the picture tube black, and the newspaper blank. They have got to produce news. If your organization has a specific objective for which you are working, then you have news. You need publicity for your organization's activities, and reporters need interesting stories packaged in a useful way. You're natural allies because you both have something the other wants.

Strategy for Getting and Keeping Media Attention at the Local Level

One of the initial tasks in developing a media strategy is to identify the media that will be effective in educating the people you are anxious to reach. If you are working on a local flood plain/wetlands zoning bylaw within your community, you do not need to reach the *New York Times*. You need to become well-acquainted with the editor of your local daily or weekly newspaper. You do this by telephoning and asking for an appointment at his or her convenience. Make certain that you arrive on time, armed with a well-written press release and a carefully organized explanation of the importance of wetlands and flood plains to your particular town and to the flood control and water resources of towns downstream. Since editors appreciate information given by experts, it is always helpful to have a quote in your press release from an expert on the subject. It is also helpful to be able to say that passage of your bylaw is endorsed by the Rotary Club, the garden club, and the local League of Women Voters. The support of these and other similar organizations gives your cause credibility. The editor pays attention, and chances are very good that this week's newspaper will carry a story on the flood plains/wetlands bylaw. Perhaps, with a suggestion from you, there may even be a picture, particularly if you have it with you and have taken it according to the paper's specifications.

Now, let us assume that your bylaw has made the newspaper; and the opposition begins to flow into the paper from real estate interests and land owners along the river who own the wetlands and flood plain that you are planning to remove from development. Because of your careful explanation to the editor, he or she will already understand the importance of the issue, and will not be easily swayed by those seeking to make a profit by filling and building on wetlands and flood plains. However, your group must carefully monitor the paper, and if you have local cable TV, you may wish to ask to do a well-prepared program on the subject. Again, remember to include names of the groups that endorse the bylaws. Send your most photogenic and well-spoken representative for television appearances. If you are not good before the TV cameras, don't go; send the person you are convinced can do the best job of convincing the public that the town should pass the bylaw.

Letters to the editor from various groups endorsing the bylaw will be of real assistance. They are a gauge of public sentiment. Don't begin sending letters too early, because letters often bring response from the other side. Be certain, if the opposition seems to be writing a great number of letters, that you more than match the number and the quality of the arguments.

If there is a public hearing on your bylaw, be sure that you have your constituency there in large numbers with copies of their statements for the press, and make certain that you and members of your coalition or group are "dressed" for the TV cameras. Remember that your purpose is to convince the public to vote for your zoning bylaw. Be sure that your arguments are not overstated, and that they are credible. At the same time, do not let the opposition get away with making statements that are convincing but untrue; and do not hesitate to call attention to any conflict of interest, such as property ownership in the area under consideration for zoning for the public welfare. However, remember to keep your control and make carefully considered statements of fact.

Strategy for Getting and Keeping Media-Attention
Issues to Be Decided by Large Constituencies

While many of the tactics are similar, the job of dealing with the media on a state-wide, regional, or national issue becomes much more complicated. To educate the public on your side of an issue such as returnable-container legislation or protection of the outer continental shelf from oil exploration is a long and complicated process, particularly since a great deal of publicity and money is being spent by the opposition.

First you must identify the media, preparing a list of newspapers (daily and weekly), radio stations, TV stations, wire services, and organizational newsletters (from churches, unions, clubs, environmental and public-interest groups, and so on). Make your list comprehensive for ready reference. Your next step is to review the list, asking yourself the following questions: Who is your audience? Whom do you wish to reach, and which stations and papers have this kind of audience? What is their geographic range, and what is their circulation? Who runs the paper or station? What is their editorial slant, and what stands have they taken? Finally, do you know anyone who could give you an introduction to any of these media?

Personal contact with the media is important. Open and honest dealings with the media will build trust. If the media know you your press release becomes more credible. Personal contact between media and your organization definitely increases the likelihood of having them publish your press release. If you know someone who knows people at the paper or TV station, ask that person to go with you to make the introductions. If there is no one, then you must go yourself armed with a well-documented press release and, if applicable, the name of the reporter covering environmental affairs. If possible, call ahead and arrange for an appointment with the environmental reporter or check to see if he or she is in the office. A personal appearance certainly gets your cause more attention.

Press Releases

The main source of information for the media is the press release. Use them liberally, and send them to everyone on your press list. The release should be brief—no more than two pages double-spaced. The most important items should be in the first two sentences, followed by those of lesser importance, with quotations and a discussion of the general issue. The first paragraph and each succeeding paragraph should be able to stand on its own; that is, the story should make sense if it ended with the first paragraph, the second, third, and so on. This enables the media to use just parts of the release easily. Quotations are always helpful. The style should be interesting, informative, clear, and concise. Present your point of view with objectivity. Remember that a story has a greater chance of being printed when very little editing is needed. The date of the release should be in the upper left-hand corner with the contact person and telephone number in the upper right-hand corner. If it continues on a second page, type (*more*) at the bottom of the first page. The journalistic note -*30*- ends the release below the last line.

If there is to be an event, the release regarding it should be hand carried (if possible) in advance, and copies should be passed out at the event. Individual packets are often made up for the press.

Press Conferences

When you plan a press conference, remember that timing and deadlines are extremely important. Get a feel for when you will draw the greatest number and most important media for your particular issue. Don't plan the press conference before 9:30 A.M. or after 2:00 P.M. It won't be on the six o'clock news, and may not be covered at all. Remember that when you hold a press conference, you are responsible for the logistics. You are giving your message to the media. The setting is important. It is better to have the room be too small than to have a few media representatives meet in a large room. An office conveys a sense of activity and competence. Be sure that you have extension cords and electrical outlets, as well as a podium for speakers. Make certain you have informed the media early enough so that they can attend; and call the media again an hour or two before the press conference to make sure they are sending someone. Don't be embarrassed about making the call to remind them; the reporter you know may have been sent to a fire, and someone else may attend after your call.

In advance of the press conference, all members of your group who are planning to speak (and even those who are not) should be briefed and in agreement with the press statement. The more competent people feel, the more effective they are. Those planning to speak should go over what they are planning to say so that it is understood and approved by the group. One person who is not planning to speak should be responsible for making sure that the media have coffee (and doughnuts if you can afford them), as well as meeting their last minute needs for materials. Do not have too many people speak, and do not be repetitive. Include in the press packet and in your presentation the names of the other organizations supporting you, and have a person there from that organization to answer potential questions as to why they are supporting you.

Begin your press conference on time. You should wait briefly for the TV cameras, if they have not yet arrived, or for a special reporter who may be late, but never delay beginning the press conference by more than ten minutes. Have someone ready to make last-minute calls to ascertain whether reporters who are not yet there are just late or are not coming because of a breaking story. Do not delay long enough to lose those who may have other commitments. After the conference ends, you can always go over details with the person who was late.

When the cameras and microphones arrive, the press conference reaches beyond those at the podium. The importance of the natural resource, and why it should be protected, becomes the message. This is your opportunity to educate the media and those at home who know little or nothing about the subject. Do not assume that they know what you know. Be sure to have some interesting visuals—charts, displays, and other evidence of your position—that the media can readily comprehend. Your organization will not get more than a one-minute spot on the evening news, so make the entire press conference count. Open by stating the purpose and introducing the participants. Give your strongest statement first, and then let each speaker present a different aspect of the issue. Close with a short summation; then open the meeting to questions. The entire press conference, including questions, should last no more than thirty minutes.

Don't be surprised if you panic shortly before the press conference with the fear that no one will come. This is a normal sensation. There are times when no one does come, particularly if an important story breaks, or the building down the street catches fire, or someone murders someone else. If no one shows, call the media and ask why they weren't there. (Be sure to carry dimes for this purpose.) If they have no reason, ask to come down to their office and discuss the issue. If this cannot be arranged, swallow your pride and hand carry your press packets to the media. Be friendly as you deliver them, and talk about the issue with those you give the release to. Sometimes that works amazingly well; and certainly you are educating someone—and who knows whom they may educate. It's all part of the process of spreading the word.

Op Eds, Weekly Columns, Community Calendars, and Letters to the Editor

Op eds are editorials by the public printed opposite the regular editorial page. The custom began with the *New York Times*, and many papers have now adopted it. Make your editorial no more than three hundred to five hundred words. Such editorials often serve two purposes: first, if printed, they educate the public; second, they educate the media. Often, after sending in such an editorial, you will hear from the media. They may even wish to interview you or endorse your position. Usually, such contributions are unsolicited, which leaves you the initiative. Never permit an op ed written by the opposition and containing statements with which you disagree go unanswered. As Dag Hammarskjold once said, "The Madman shouted in the Marketplace. No one stopped to answer him. Thus it was confirmed that his thesis was incontrovertible."

Many local newspapers have weekly columns open to their readers. This is an invitation for you to educate the people in the community. These columns are read and often do an important educational job. However, they do require a commitment to write a column once a week. If this does not seem possible for you, ask the newspapers if you can write a guest column for one issue, or write a column once a month. Getting to know the local newspaper editor makes all things possible.

Almost all newspapers have a community calendar. By listing your events, you attract a broader constituency and bring attention to your project. Remember to meet the community-calendar deadlines. Don't send the information too early—it will get lost, and don't send it too late.

Letters to the editor have been discussed earlier. However, the importance of this tool cannot be overemphasized. The letter page in every newspaper is the most accessible. Most newspapers have a commitment to printing a selection of all letters received. The letters are the so-called voice of the people. In organizing your letter-writing campaign, timing is extremely important. Try to organize it so that you have the last word. Letters often bring responses from the opposition, so you must be prepared to have a network of persons ready to respond to these letters in an intelligent, well-organized manner. In weekly newspapers, it is possible to time your letters and/or appearances so that they will be read just prior to the vote on the issue by the decision makers.

Public-Service Announcements, Personal
Interviews, and Telephone Interviews

Just as newspapers have community calendars, radio and television stations have public service announcements (PSAs). However, they differ from the community calendar in that they can also be used to broadcast an opinion on controversial issues. Television stations do not like to do this, but the Fairness Doctrine mandates that they broadcast the issues. Directors of stations do not want to have complaints lodged against them. Knowledge of this makes it easier for you to publicize your organization's point of view. A PSA should be submitted in script to be read by a media person or on a tape made by the most well-spoken member of your group. PSAs should conform to station requirements, which usually require submittal several weeks before broadcast. They should be timed to fill the allotted slot, which is usually thirty to sixty seconds.

The personal interview and the telephone interview are two other ways to command media attention. Most stations have regular interview shows or allot space to interview community representatives on current issues. If they do not invite your group or coalition to speak, call and ask them if you can come. When you have an opportunity for such an interview, be sure to arrive at the station early. You may have an opportunity to visit with the person who will be interviewing you and point out the areas of concern that you wish to discuss. If the interview is going to be held with pros and cons being presented, you must send the person who recently left the college debating team—the person who thinks fastest under pressure. Do not, even if you are the motivator for the entire project, accept such an interview if you do not have the experience and expertise to take on the professionals smoothly. Send your most professional speaker, making sure that person is extremely well informed on all sides of the arguments before leaving for the station. Brief your representative carefully. If he or she is able to present your arguments clearly and convincingly while shooting gaping holes in your opponents' statements, you have won one educational round.

A personal appearance at a station also offers an opportunity to make a taped interview. If you hand deliver a press release, a public-service announcement, or a backgrounder (a paper providing information on a current topic), you must might be asked to discuss the issue on which you are currently working.

Another common form of interview takes place over the telephone. The first time you are called by phone and asked to be taped, it might be wise to request to call back in a few minutes. Just hanging up the phone and thinking for a few moments will enable you to collect your thoughts and organize your responses in such a manner as to make maximum use of the time you will be given. Remember always that you are speaking to the public; and most of them know very little about the topic that concerns you. Go back to the basics if possible—give the listener, in lay terms, the reason for your activity or concern.

Part of the planning for any event, whether it be a demonstration, a workshop, or a conference, should include publicity for the event. An "advance" is a short, double-spaced, one-paragraph description of the planned event, clearly including date, time, place, major speakers, and groups co-sponsoring the event. The headline on the advance should be written so as to command the attention of the media. The advance should be typed on the organization's stationery and should include the name and phone number of the contact person. Mail the advance so it arrives one or two days before the event takes place. If you are having a large event, send the first advance a week and a half ahead of time, with a follow-up one or two days before the event. The person hand delivering an advance may be given an opportunity to address the subject—another educational opportunity.

Press calls follow the advances. Persistent frequency is the only reminder of your event. The media are receiving hundreds of advances; your job is to get them to your event. Reporters are constantly changing. You must call both the night before and the day of the event to remind those reporters on duty. This is particularly important for weekend affairs because weekend personnel vary from those on the normal schedule. If you have speakers, press packets should contain copies of as many of the speeches as possible. You do not have to pay for luncheon or dinner for the press if you are unable to do so. You can and should, however, offer them free admission and suggest to them a restaurant close by, or give a reduced price for luncheon with the group. This depends on the financial status of your organization.

No chapter on the media would be complete without some discussion of the Fairness Doctrine. Unlike newspapers, which, theoretically, anyone can start, electronic media have available only a limited number of airways—too few to meet the demands of the people who want to use them. The statute that governs the use of radio and TV endorses the doctrine that the "airwaves belong to the people." Stations must operate for the "public convenience, interest, and necessity." According to the Federal Communications Commission (FCC), the controlling and licensing agency for both radio and television, stations are stewards on the public behalf. If stations do not serve the public interest, they may lose their license.

While serving the public interest is a broad mandate, there are two main guidelines. One of these is the Fairness Doctrine: Stations must air a balance of opinions on subjects of community interest, presenting both sides of the issue. That means that in order to present a balanced view, station time must be made available to those who do not necessarily agree with the station's editorial commitment. The station has the right and the responsibility to determine whether they have presented a balanced view. Although it may not air your particular request, it must cover your side of a question if the other side is being presented—even if the other side is paying for advertising. If you feel that you are being treated unfairly, you may lodge a complaint with the FCC. This was successfully done in California when pronuclear utilities put on an advertising campaign against a nuclear-power referendum. The second guideline requires that stations seek out and present issues of public importance. The obligation is not as specific as the guidelines governing equal time. However, it does bestow on stations a responsibility, which can be used to convince directors of your subject's importance.

It is important that environmental organizations learn to use the media wisely. The FCC, with its Fairness Doctrine, is responsible for keeping the coverage of important issues before the public and making it possible for citizens to hear all sides of a question. This is sometimes hard to do when advertisers are paying tremendous sums to the television and radio stations and to the press. Volume 39, Number 139 of the *Federal Register*, July 18, 1974 explains the Fairness Doctrine for electronic media. It would be wise for environmental organizations to review it every so often, particularly when they hear the "other side" of an issue receiving more attention than they feel it should receive. Remember, as Marshall McLuhan said, "the media is the message," and certainly, in a democratic society, the public interest must be protected.

11

Cable Television: A Tool for the Environmental Movement

Mary Beth Bablitch and
Stuart Langton

Americans are only beginning to realize the significance of the communication revolution taking place in society. As terms like *videotext, satellite, link-up, software* and *hardware* become a part of our everyday vocabulary, our businesses, government, and educational institutions are preparing for unprecedented communication opportunities. For environmentalists and members of other public interest groups to remain effective in the future, they must also understand and learn how to take advantage of the new ways of obtaining and communicating information.

One of the new communication media environmentalists least appreciate is cable television. At present, many of our cities and towns have recently developed or will soon grant franchise agreements to have cable television systems installed in their communities. The great promise of cable television is that it offers citizens a wide number of channels to view at an affordable cost. Further, cable TV provides a variety of new services and communications opportunities.

For example, viewers in a number of cities and towns are already shopping, making plane reservations, checking the stock market, and even banking via cable television. Many houses across the country will soon be linking their TV screens to data-information banks, video newspapers, audience-polling devices, and retail mail outlets. While over 27 percent of American households are now cable subscribers, many analysts project a cable-penetration rate of over 50 percent by 1985.

Cable TV and Environmentalists

Environmental activists and representatives of other public interest groups may see cable television as a boon, or just more of the same TV. The difference will depend upon the capacity of these organizations to use the media for purposes of education and advocacy. Their abilities to respond to the challenges presented by all the new communication technologies will relate directly to their understanding of the implications of the communication revolution, their anticipation of community-cable-television's impact at the

119

local level, and their willingness to become involved in the practical aspects of community cable television.

A New View of Television

Traditional commercial television has alienated many people because of its severely limited program selection and the lack of community control over content. These feelings have inspired the use of terms such as *wasteland* and *the boob tube* to describe the medium. The use of cable TV and related technologies has the potential of altering these perceptions and ushering in profound changes of how people will view television in the future.

Cable television, computers, telephones, and satellites (technologies once thought of as independent), when combined, yield remarkable new information-delivery systems. Video recorders and inexpensive home video production systems will also play an increasingly important role in our personal lives, homes, and social institutions. Although this may sound like Buck Rogers-ese to many people, these cable-related technologies have the potential of transcending the traditional patterns of alienation in televiewing. Four of these ways are particularly important to examine in the context of community cable television today: increased program selection, planned viewing, interactive capability, and community education.

Increased Program Selection

One of the most unique aspects of cable television is its potential for increased channel capacity, with up to one hundred channels having been allocated in some cities. The development of this increased channel capacity has spawned the idea of narrowcasting: programming tailored to the interests of specific groups in the population. Eventually, hundreds of groups could be served by their own cable-TV channel, resulting in a video smorgasbord of sharply focused issues, problems, and concerns. Such cable programming may provide the kind of opportunities in the next twenty years that specialty magazines have provided in the last twenty. In this respect, it is not unrealistic to expect that, in some areas of the country, a special channel could be devoted to environmental issues.

Planned Viewing

The number of channels and variety of programs available through cable systems offer great opportunities for people to develop a much more planned and discriminating attitude toward television. In this respect, television may come to be seen less as a diversion and more as a resource. Increasingly, people will turn to television to select programs that address their interests, instead of just wondering, "What's on now?"

The diversity of programming available through cable television will have an even greater effect when coupled with the use of video-recording devices. Instead of adapting to broadcast schedules, people will be able to record programs of interest and view them when convenient. This will enable people to plan their time more consciously, and it will encourage greater individual awareness of all television broadcasts during the week, rather than when viewers have free time.

Interactive Capability

Cable TV provides the viewer with a new relationship to television because the viewer is able to provide feedback to programs. Using a home-control console, viewers can express their preferences on issues, indicate the strength of their feeling on a public-policy question, or indicate their level of interest in proposed programs. In this sense, interactive cable brings a new form of democratic participation into the living room.

Community Education

A wide range of institutions and individuals are likely to be drawn into television programming. Educational institutions, nonprofit groups, self-help organizations, lobbyists, and technical experts may find that specialized cable channels provide outlets for their needs and interests. These types of opportunities may, in turn, become a powerful new means of education and citizen involvement in a community.

At the same time, video cameras will become more important to citizens and community institutions as local cable companies set aside channels for local citizens to present programs that they produce. As these local-access channels, as they are known, grow in importance, people will want to record messages and programs for other community members. As the cost of video cameras continues to decline, they will increasingly become a new communication tool for the concerned citizen, government agency, or non-profit group.

The current challenge to environmental organizations is to recognize the force of these changes. Media strategies and staffing patterns should be adjusted accordingly. As television-program production becomes more accessible to the public and programs become more specialized, our future vision of the educated person and competent organization will be expanded to include knowledge, values, and skills related to the new television and associated electronic technologies.

The Community Impact

The full potential of cable television can only be realized by considering its impact at the local level. In this regard it is important to distinguish the community cable system from the remainder of the cable operation.

This distinction can be understood best by describing three of the most unique elements of community cable television which are popularly referred to by the terms *local origination, public access,* and *two-way-interactive television.*

Local-origination programming is initiated and controlled by the local cable company itself. Typically, one or more staff persons are assigned to produce or procure a given amount of local programming on issues they see as being important to the community. There are two very different sources for this programming. The first involves tapes that have been produced outside the community and are purchased or rented by the local cable company. This outside taped programming is considered "local origination" in the sense that the local company decides what specific programs to telecast and when and how often to telecast them, based upon the perceived community interest in them. The second type of local origination is produced within the community and draws upon the community for the programming material. It can be telecast from the cable-company studios or from the portable equipment elsewhere in the community, and it can be shown live or taped and edited for subsequent use on the cable system. Whichever the source, local-origination programming tends to be the most professional, but also the most independent from community involvement, of the forms of community cable television.

Public-access programming is initiated by local organizations and individuals outside the cable company. The key difference between public access and local origination is community involvement. In its purest form, community members perform all the work necessary to produce a public-access program. At the opposite extreme, an individual's involvement may be limited to a request to the cable company's public-access coordinator that his or her organization's event or issue be videotaped with the cable-television equipment. Typically, however, volunteers play a key role in developing and producing the programs, with significant training and advice from cable-company personnel. In most such cases, the cable company also provides all necessary facilities and equipment for the production and transmission of the public-access program. The result may be less polished than a network production, but it tends to serve better the feelings and needs of the community. In addition, public-access efforts have the potential to produce fresh and creative programming unequaled by any other source.

Two-way-interactive programming allows each cable subscriber who has the necessary equipment to send back a signal over the cable system from his or her own household to the central studios. The unique aspect of two-way cable television is the range of services it can provide: banking from home, ordering groceries via cable, interactive information and educational services, home fire and burglary alarms, and many more. In each case, these services involve sending a message from the cable subscriber to the appropriate business, municipal service, or agency by route of a trunk line from the cable-company central facilities.

In addition, the two-way communication capability can be used in combination with either the public-access or local-origination programming discussed above. A local citizen group could develop and produce two-way programming to encourage community participation in their concerns. Alternatively, the local cable company could develop and produce two-way programming for the community. Such interactive production has the potential to allow more citizens both to be a part of community decisions and to better understand the perspectives of fellow community members.

Through community cable television, citizens can create programs that capture the unique spirit of a city or town. Local organizations can, for the first time, use television to communicate effectively to audiences they most need to reach. And local events can be reported in detail rather than being ignored by larger metropolitan television stations.

Environmental groups stand to benefit from community cable television in many ways. The use of video to get information out about specific environmental issues can be a significant addition to many organizations' educational and promotional efforts. The impact of a visual medium can go far beyond the printed word or the audio presentations of radio. It can attract more people and retain their interest longer in discussions of complex issues. A video presentation can graphically demonstrate the nature of problems and increase people's understanding of alternatives.

Cable programs that present timely information about environmental legislation, administrative changes, and views of government officials can stimulate widespread discussion of issues that are complex and controversial.

Whether the format be a debate, a meet-the-press forum, question-and-answer programs, panel discussions, interviews, or talk shows, much can be gained by using a communication strategy that offers a visual dimension. Environmentalists can better communicate their message to wider audiences through these cable-program productions, while still utilizing local newspapers, radio, and broadcast networks. In turn, the electorate can be expected to broaden its perception of environmental issues.

Becoming Involved in Community Cable Television

Much must be done to ensure the best and most appropriate cable-TV system for a community. Before the benefits of community-cable-program production can be realized, proper anticipation and attention must be given by local communities to two critical areas: franchise selection and planning; and governance of community-cable-programming decision making. Citizen involvement in these two areas can help guarantee that the needs and interests of the community are reflected in the overall design of the community-cable-television system. Environmental organizations are also likely to see policy adopted that fits their needs if they are involved at the initial stages, as well as throughout the life of the system's development.

Franchise Selection and Planning

A franchise, which is granted by local government to a cable-television company, authorizes the construction and operation of a cable system in a community. The type of franchise that is arranged between a community and a cable operator should be specifically designed to meet the needs and interests of the community. Since most franchise agreements are granted for ten to fifteen years, it is important to consider all possible options.

As part of the franchise agreement, a cable operator is usually required to make a substantial commitment of resources to support local cable programming. Before this happens, it is especially important for community residents to express their community-cable interests and influence the access packages proposed by the competing cable companies in their bids for the contract. A community's cable needs can be determined, in part, by talking to members of local organizations. They can provide information on which groups are likely to utilize a public-access channel for the purpose of producing programs. This kind of information will help define how many access channels are needed and how they will be utilized, (that is, how many government, educational, public, and two-way channels should be set aside). In addition, consideration should be given to general equipment requirements, facilities, staff, and budget-line items for publicity, videotape, equipment maintenance, and the like.

Currently, much is being written about communities' experiences with cable-television franchising. Learning from these experiences will help other cities and towns shape the best possible contractual agreement. Participation in regional and national organizations that provide information on community-cable-television developments should also be considered. Likewise, involvement in cable conferences and seminars will help provide information on state-of-the-art developments in community cable television. Ultimately, the considerations stated in the franchise agreement will help establish minimum and maximum service requirements and expectations from the firm to be awarded the franchise. The more specific a community is in its final demands, the more likely it is to get what it wants from the cable operator.

Governance of Community-Cable-Programming Decision Making

One of the most significant ways community members can become involved in the acitivities of planning and developing a community cable system *after* the franchising process is by participating in the actual governance of the system. An active and meaningful citizen participation process at this level is the key to effective community cable television. As communities throughout the country have learned, the potential of cable for addressing community needs can only be realized by incorporating citizen interests in the decision-making process.

Citizens interested in the promotion of environmental issues and concerns can make sure their voices are heard by taking part in the overall governance of their local cable system. The decisions made at this level encompass the important elements of local origination, public access, and two-way programming. Issues ranging from defining the programming needs of the community, stimulating public access, and achieving a balance between what subscribers want to watch versus what residents are able to produce, are a few examples of the concerns citizen-advisory groups address. In addition, practical evaluation of equipment, training priorities, use of volunteers, and methods of outreach to those who might produce local programs are addressed.

To be effective, a citizen-participation structure should be developed as quickly as possible after a franchise is granted; and it should provide procedures for local citizens to establish policy and implement day-to-day decisions involving public-access, local-origination, and two-way programming on cable television. Such a citizen-participation structure could become empowered by the local franchising authority or the operating cable company.

The formation of a well-defined participation process can help determine and implement responsible community-cable policy. However, the exact type of citizen-participation process required in one community is not likely to address adequately the needs of the next. Despite differences in communities, each should seek to develop a participatory structure that will serve certain goals. Specifically, the citizen-participation structure should: (1) be *representative of the community itself*; (2) be *responsive to cable subscribers*; (3) be *responsive to cable activists*; (4) make the *best possible use of volunteer resources: time, energy, and expertise*; (5) assist the cable system to achieve *quality programming*; and (6) ultimately achieve and maintain adequate *citizen support and backing*.

Each of these six goals directly competes with the others. Most systems lack the resources and staff power necessary to achieve all of them completely. The art of deciding what balance to seek, and how to achieve that balance, is an important part of any community-cable operation. On the other hand, an excessive concentration on just one or two of these goals, to the exclusion of others, could significantly weaken the system. In order to approach the full potential of the community-cable system, all six goals should be systematically addressed by a representative group of concerned citizens, and leaders of environmental groups should become actively involved in this process since they have much to benefit from it.

Seven Tips for Effective Use of Community Cable Television by Environmental Groups

This essay has suggested the growing importance of cable television on local communities and its potential for environmental education and advocacy. To conclude this discussion on a very pragmatic note, we further suggest that there are seven specific activities that will help environmental organizations take advantage of this new communications medium.

1. As outlined earlier, environmental organizations and their leaders should become involved in the franchising process. This can involve lobbying town or city officials to identify environmental programming, as well as involvement in determining the criteria used to select a cable company. Environmental leaders should also encourage the selection of a company that makes a strong commitment to local access, community-developed programs, good production facilities and technical assistance for developing local programs, and substantial training and community participation.
2. Environmental organizations should actively participate in the governance of the local community-cable system. This can include placing leaders of environmental groups on advisory committees, attending meetings of such groups, and being placed on their mailing list. On an ongoing basis, environmental groups can offer suggestions regarding relevant environmental issues and programs that might be addressed on

community channels. However, in so doing, environmentalists should realize that resources available for producing local programs will be limited. Therefore, they should also be prepared to volunteer in the production of local programs, as well as in fund-raising events that help supplement the funds a cable operator provides for community programming.

3. Environmental groups should consider the use of cable television as one of the ongoing communication media they utilize. To this end, they should become acquainted with the staff of the local cable station. As they engage in annual planning or planning for special projects or campaigns, they should always question how they can best use cable television to serve their goals.

4. As environmental groups engage in fund raising, they should consider using cable programming as a special item to appeal for funds and equipment. Video recorders, videotapes, and tape-splicing equipment are among the useful equipment they might request. Requests for assistance in producing special programs on critical local environmental problems may be appealing to some funding sources.

5. To make maximum use of cable television, environmental groups should engage in training for their staff and interested volunteers in how to produce local programs. Many local cable companies provide such training as a service to nonprofit groups. Where such training is unavailable, several environmental groups or a consortium of nonprofit organizations can arrange for their own training through a local college or university. Once a number of local leaders receive training and become experienced, they can conduct training seminars every several months for other interested community leaders or groups.

6. In order to keep local citizens informed of environmental issues on an ongoing basis, one or several environmental groups can establish a committee on environmental news. Such a group can provide weekly reports for the local cable-news program. They might also provide a person who can serve as a regular environmental commentator on a local news program. Another possibility is for the local committee to produce a weekly environmental news program.

7. Local environmental groups should consider producing a number of special programs each year on environmental issues critical to the community. In developing such programs, a number of suggestions are in order. One is to err on the side of producing fewer programs of high quality rather than many programs of mediocre or poor quality. In the long run this will provide more credibility and greater interest. Also, it is helpful to produce programs that are of interest to the majority of the community, rather than ones dealing only with issues of concern to environmental leaders. To identify such issues, surveys of a random

sampling of the population can be conducted on a yearly or biyearly basis. In this way, environmental organizations will reinforce their image of being concerned about community interests as a whole, rather than with narrow or fringe issues. Finally, for political reasons, consideration should be given to program scheduling. In particular, programs designed to influence voters or elected officials should be shown shortly before critical votes on these issues.

12 Networking and the Environmental Movement

Stuart Langton

In recent years, *networking* has become an extremely popular term. Although networking has long been a specialized area of study among anthropologists and ethnographers, the notion has become increasingly fashionable in our institutional life in the United States. For example, a number of groups, such as the Gray Panthers, use the title "Network" for their newsletters. Many women's magazines describe the value of networking on the job. One company has an organization called "The Network." The topic is popular in many professional conferences and journals, and there is even an organization created to study and advocate networking.

An examination of the different ways in which the term networking is used suggests that it has three separate meanings. First, it is a *technical concept* used among anthropologists to describe and measure influence patterns in human relationships. Second, networking is often used as a *general metaphor* to represent a particular kind of helpful relationship that can be created between individuals. Third, networking is increasingly viewed as an *organizational strategy* to serve the interests of different groups and their members more efficiently.

Networking in all three of these senses is a very meaningful idea for the environmental movement. In our increasingly complex and bureaucratized world, networks represent quick and effective ways of moving beyond our institutional boundaries to share information and accomplish things. "Networks," suggest Jessica Lipnack and Jeffrey Stamps, "are the lines of communication, the alternative express highways that people use to get things done. In crisis or opportunity, the word spreads quickly through these people-power lines." They continue, "Networks and networking are the structures and processes through which the ideas and values come alive."[1]

Network processes and structures are particularly relevant to environmental groups for three reasons. One is that few environmental groups have sufficient resources to achieve their goals without a lot of help from others. Another reason is that many issues of concern to environmental groups are of a crisis nature. Therefore, environmentalists must mobilize information, resources, and support very quickly—and networking practices are the best way to do this. A final reason is that environmental problem solving requires a great deal of institutional interdependence. In addition to working together for the reasons cited above, environmental groups must relate to many government agencies, businesses, educational institutions, and the media. Informal networking contacts with like-minded people in these institutions is frequently the best way of avoiding the inherent bureaucratic hassles and barriers we encounter when we want to get something from one of these institutions.

The purpose of this essay is to encourage environmental organizations to adopt networking practices as a conscious organizational strategy. However, to do this effectively, it is helpful to understand some of the findings of anthropologists over the past thirty years; these findings are the result of methods they refer to as *network analysis.* In addition, the powerful potential of networking is better appreciated if we consider why the term has become such a popular metaphor.

Some Learnings from Network Analysis

The field of network analysis, which began in the 1940s, is principally concerned with describing and measuring patterns of influence in human relationships. The basic thesis of anthropologists in analyzing networks is that it is possible to determine how change takes place in a community. J.A. Barnes, in a pioneering study of a Norwegian island community in 1954, used the image of points to represent people, which were then connected by lines as a way of graphically charting who interacted with whom.[2] Since the work of Barnes, the field has spawned a host of technical terms, graphic methods, and mathematical models. Nonetheless, network analysis as a field suggests a number of very clear and basic principles for environmentalists.

First, the practice of analyzing networks can benefit environmentalists or any public-interest activists who would like to exert influence on a community. In this regard, it is a helpful tool for planning educational or advocacy campaigns. Taking the time to analyze a network helps identify pivotal persons in a community, many of whom would not be recognized otherwise and who can be critical in influencing others.

Second, network analysis has made it clear that there are all types of networks, networks are present everywhere, and individuals may belong to a number of different networks at one time. Some of these networks may be partial, or limited in intensity, and others may involve people in a more total way. So, for example, a family, a neighborhood, a professional association, or even an area of the country might be viewed as a network of interacting persons. Of significance to environmental leaders is the fact that networks may emerge in a spontaneous way without members consciously realizing the nature of their network, or a group of people can consciously choose to create a network which serves some clear-cut objectives. Most environmentalists already do network in some form or another, and developing other, more formal, networks is something leaders may choose to do.

Third, networks are not merely patterns of relationship between individuals, but also include linkages between organizations. In addition, institutions can consciously decide to network by sharing information or cooperating in a common activity. This seems to be happening increasingly among environmental organizations. For example, since the early 1970s, environmental organizations in New Hampshire have formed a state environmental coalition to select priority legislation on which they would work together. Since 1977, the Lincoln Filene Center for Citizenship and Public Affairs, at Tufts University, has coordinated the New England Leadership Network to provide training and share information among organizations within the region.

Fourth, the nature of the connections that can be made between people or groups in a network may vary. The connections can include exchanges of information, goods, services, money, advice, or political support. Furthermore, exchanges in networking do not have to be of the same kind. For example, one group may give another goods in exchange for political support. The exchange does not have to be equal; this should be instructive to environmental leaders in organizing networks, in that they should consider what it is they have to offer and what it is they might receive from other organizations.

Fifth, as Jeremy Boissevain points out, networking is facilitated by certain types of individuals with particular skills for the task. Referring to such persons as social catalysts, brokers, or network specialists, Boissevain concludes that they "provide important links in networks viewed as communication channels. They transmit, direct, filter, receive, code, decode, and interpret messages."[3] While some persons perform these functions naturally, there is no reason why these skills cannot be taught and learned by those who might apply them on behalf of an organization. To go a step further, it would be in the interest of environmental organizations to encourage the development of such skills among their leaders.

Network as a Metaphor

The growing popularity of the idea of networking raises several interesting questions: Why is this notion so popular? Is it merely a fad, or does it represent something deeper?

I propose that it does represent something deeper that is conveyed in the general, metaphorical use of the term. As Ernst Cassierer once pointed out, metaphors, like myths and works of art, usually reveal deeper feelings of fear or aspiration among people.[4] The somewhat vague admonition *to net-work* seems to reflect a unique and widespread aspiration of a great many people whose identities are strongly rooted in their careers and whose energies are devoted to institutions they serve (as professionals or volunteers).

What the notion of "network" as a metaphor represents to many people is the desire to experience a sense of community in a time when more traditional forms of community are eroding. Whatever else a "network" may mean, at a very existential level it symbolizes an extended community, beyond geographical boundaries, of people who share similar values and are willing to share and assist one another.

It is the desire to discover and relate to like-minded persons in helping ways, I believe, that makes the idea of networking so attractive to people. In many respects, it is an antidote to the experiences of depersonalization and estrangement that we so frequently experience in our contemporary world. As family traditions, community institutions, and common social values continue to weaken, networks suggest one positive form of new relationships that are meaningful to people. And, as our institutions become more specialized and, in many cases, isolated, networking becomes a force joining them together to serve more general and shared social values.

The relevance of this point for the environmental movement, it seems to me, is that there is probably a considerable need and desire among environmental leaders for networking experiences. The very nature of environmental advocacy and education frequently leads to battle fatigue, personal isolation, and burnout. While any healthy environmental leader needs to have a support system in his or her personal life, networking can further expand that system for purposes of fellowship, learning, and/or mutual assistance.

Three Types of Environmental Networks

While environmental leaders can create networks for many different purposes, there are three types of networks that are particularly helpful to environmental organizations today. These are advocacy networks, exchange networks, and training networks.

An *advocacy* network consists of a group of people or institutions joining together to see that a public policy is enacted. Although the target of such networks is usually government, it may also include businesses. Advocacy networks usually evolve through several stages, including: the shaping of a common political agenda; the selection of specific proposals to advocate; and the sharing or coordinating of resources in advocating the proposal. An example of an advocacy network is the OSHA Environmental Network, which drew together a number of national environmental groups and labor unions to advocate various types of worker safety and environmental legislation. Another example is the New Hampshire Environmental Coalition, mentioned earlier.

Basically, an advocacy network is a way for an environmental group to increase its political clout by joining forces with other organizations. This is a very much needed practice for several reasons. One is that our political system has been overwhelmed by interest group lobbying. Direct lobbying by several special-interest groups has had the effect of diluting each group's influence because of the sheer volume of groups. Consequently, the larger a network, the greater its potential for impact. Another reason why advocacy networking is needed is the growth of legislative and regulatory issues that government must deal with due to technology and the complex nature of our society. Because there are so many environmental issues floating around at all levels of government, no single environmental group can keep track of or respond to them all. An advocacy network is a more efficient way for environmental groups to share responsibilities for identifying, studying, and responding to the host of environmental issues at each level of government.

An *exchange* network consists of a joining together among persons and/or institutions to assist one another. Seymour Sarason, Elizabeth Lorentz, and their colleagues, who have written insightfully about the dynamics of such networks, refer to such interactive associations as "barter exchange networks."[5] The purpose of such networks is to benefit each party in the network through the sharing of resources. Since the resources that can be shared may be as different as money, office space, staff assistance, photocopying or printing, supplies, telephone use, mailing lists, and advertising, the members of an exchange network "barter" with one another to determine what resources each needs and what resources each has to give.

As Sarason and Lorentz observe, sharing resources in this way is of great practical value today because few organizations have or will have all the resources they need to achieve their goals.[6] This is certainly true of environmental groups, which always seem to be operating with limited resources. Unfortunately, as Sarason and his colleagues also note, "In everyday practice, agencies do not seek each other out for the purposes of resource exchange; each agency sees itself as independent of all others, dependent only on its own subsidized resources . . . and nurturing the fantasy that there must or there should exist the quantity and quality of resources that could ensure a safe and goal-fulfilling life."[7]

A *training* network is the third form of networking that may be of considerable benefit to environmental groups. The purpose of a training network is to create learning opportunities for leaders of participating organizations to develop or expand their knowledge and skills. To create such a network usually involves three basic steps. First, the groups need to conduct some type of needs-and-interest assessment among their leaders to detemine the shared subjects they want to address. Second, the network members need to agree on what procedures will be used to provide resources for the network. For example, the groups might pay a general fee; each group could accept responsibility for conducting a different training event; or each group might pledge to send a certain number of paid participants to training events that would be coordinated by one or several of the network organizations. Third, the network members need to decide on the most appropriate types of training methods. These might include courses, seminars, written materials, or technical-assistance consultations and clinics.

Training networks are a particularly good idea for environmental groups. One reason is that there is a great need for training among environmental leaders, but few groups have or are willing to invest resources for it. Not only do environmental leaders need much more management training than they currently get, but the ever-changing nature of environmental problems requires that leaders continuously expand their scientific and political knowledge. The financial pressures experienced by almost all environmental groups usually lead them to consider training as a luxury they cannot afford. This is a shortsighted view that fails to appreciate the fact that good leadership training can lead to cost savings due to better management, improved fund raising, and more effective educational and advocacy practices. Despite the value of training, few environmental groups are willing to make adequate investments of time and money in it. A training network can help overcome this problem in that the cost can be substantially reduced by spreading it across the participating groups. Further, the collaborative planning that a training network requires can produce a more attractive series of training events that are directly related to the needs of network members.

Some Principles of Networking

To those environmental leaders who believe in the value of networking for the purposes of advocacy, resource exchange, or training, the question arises: What are some of the things that can be done to create and sustain a successful network? Perhaps the best answer is that there is no single formula for success to fit all situations. However, adopting an attitude of openness to learn with others, while being experimental and creative, is probably a good place to start. Beyond that, I would like to share seven principles that I have found helpful in my own experience in helping to develop the New England Environmental Leadership Network, which over a period of six years has held eight training institutes for environmental leaders, undertaken a needs-and-interest survey among twelve hundred environmental leaders, sponsored eighteen conferences on different environmental issues in the six New England states, conducted five annual New England Environmental conferences cosponsored by over one hundred sixty groups and attended by nearly a thousand people, published a quarterly newsletter and five training pamphlets, sponsored a New England Collegiate Environmental Conference, and conducted quarterly meetings between regional EPA officials and environmental leaders.

Networking Doesn't Take Place
without Facilitators

This is one of the major points of the two works on resource exchange networks by Sarason and Lorentz, and others mentioned earlier. Further, this has been a basic maxim of the community-organization movement since the time of Saul Alinsky. I have personally discovered how important this is in creating the New England Environmental Leadership Network, which would not have happened unless my colleague, Nancy W. Anderson, was available to devote six years of working nearly full time on it.

However, although facilitators are essential to foster networking, this does not mean that they do all the work of the network. In fact, as Sarason and Lorentz point out, "Success or failure . . . is determined in large part by the leader's adherence to the logic of participation, because . . . issues of tactics and strategy should not be decided by the leader alone."[8] Further, as they point out, the founder of the successful Essex Network, in Connecticut, had to help the new facilitator learn "to stop the instinct to do things yourself." She adds, "I had not only to keep Richard from doing things, I also had to stop members from expecting him to do things." When there was something that needed to be done, I had to keep saying, " 'Your job is not to do things, your job is to think: Whom do we know, or how can we think of who could profit from doing this? For whom is this an opportunity?' "[9]

Shared Values Bind People Together
in Networking

By this, I do not mean that a catechism of moral principles must be established among those who create a network. However, it is critical to realize that you share a sense of the way things could or should be, and you feel that there are common values that bind you. This does not happen easily. It takes time to talk and, above all, to listen to one another. Sometimes this is frustrating, and this is where a facilitator can be very important. Ultimately, you must feel and understand the values and goals that make your commitment to networking worth pursuing.

Networking Must Offer Tangible Benefits

While those who network must share some basic beliefs, the network must offer tangible, practical benefits to each partner. Everybody needs to get something out of the arrangement, although what each receives may be very different. This was particularly evident to my colleagues and me in getting a working group of leaders from seventy environmental organizations to develop the New England Environmental Leadership Network. While many were reluctant to join us at first, we found that the development of three tangible benefits did attract people. First, we were able to provide leadership-training conferences at no cost to them; second, we shared some grant money we had received to support conferences organized by network teams in each state; and third, with the help of each state team, we conducted the aforementioned needs-and-interest survey among twelve hundred leaders from throughout the region and provided leaders in each state with the survey returns from their state. Eventually, this led to their conducting eighteen conferences and training programs throughout New England that responded to the needs the survey identified.

An Early Success Energizes the
Networking Process

This principle is closely akin to Saul Alinsky's rule that an early initial victory is critical for a community-based citizen organization to succeed. Alinsky's advice was to achieve an easily attainable goal at the outset. Such an initial experience of success encourages people; it gives them a sense of empowerment; it creates confidence in their joint efforts; it brings people back; and it attracts new people.

Trust Is Essential in Networking

Above all, participants must trust in the integrity of the facilitator and other leaders of the process. Network leaders must operate under the principle of "do what you say, and say what you do." Network leaders must never allow their personal interests and agendas to undermine the interests of the network effort as a whole or of any participating group or individual. Their word must be trusted, and they must communicate openly and frequently about all issues that are directly or potentially relevant to the network and its members. While disagreements and conflict are necessary and inevitable in a healthy network experience, competition among groups (in seeking funding, for example) should be avoided or should be discussed openly and be guided by some cooperative principles.

A Network Requires Continued Maintenance

This need has been described very adequately in Sarason and Lorentz's reporting of the leadership efforts of Mrs. Dewar's initial leadership of the Essex Network. What struck me in reading their account was the amount of care and time that Mrs. Dewar spent in network meetings, in talking individually with representatives, and in sending them appropriate written materials. This description closely parallels that of Nancy Anderson and me in developing the New England Environmental Leadership Network. In fact, our telephone bill became so excessive in the first three months that we had to install a WATS line. The need for written communication became so important that we created a newsletter.

Don't Romanticize or Over-Analyze
the Idea of Networking

This final suggestion may sound a little strange; but it is a principle we learned the hard way in our environmental network. When we began our effort, we read much of the literature in the field, invited leaders from successful networks to address our group, and then attempted to discuss how and why we might network and what principles we should adopt in the process. After a few days of this, it was clear that everyone was tired of talking about networking. Finally, one participant blurted out, "The hell with talking about networking, let's forget the damn word and just talk about what we want to do together and how we are going to do it." To the cheers of the rest of the group, we took his advice. While seldom referring to networking, and never referring to other models or experiences of networking, we proceeded to create our own unique and successful model.

Notes

1. Jessica Lipnack and Jeffrey Stamps, *Networking: The First Report and Directory* (New York: Doubleday and Company, 1982), p. 1.

2. J.A. Barnes, "Class and Committees in a Norwegian Island Parish," *Human Relations* (1954), Vol. 7.

3. Jeremy Boissevain, "The Place of Non-Groups in the Social Sciences," *Man* (1968), Vol. 3.

4. Ernst Cassierer, *Language and Myth.* (New York: Dover Books, 1946).

5. Seymour Sarason, Charles Carol, Kenneth Maton, Saul Cohen, and Elizabeth Lorentz, *Human Services and Resource Networks* (San Francisco: Jossey-Bass Publishers, 1977), p. 21.

6. Seymour Sarason and Elizabeth Lorentz, *The Challenge of the Resource Exchange Network* (San Francisco: Jossey-Bass Publishers, 1979), p. 14f.

7. Sarason, et al., *Human Services*, p. 23.

8. Sarason, et al., *Challenge*, p. 145.

9. Ibid.

About the Contributors

Douglas James Amy is an assistant professor of politics at Mount Holyoke and former assistant professor at Oberlin College.

Nancy Wilson Anderson is director of environmental affairs of the Lincoln Filene Center for Citizenship and Public Affairs, Tufts University, and director of the New England Environmental Leadership Network. She is a former president of the Massachusetts Association of Conservation Commissions.

Mary Beth Bablitch consults with communities on community cable television concerns. She is a former staff associate of the Lincoln Filene Center for Citizenship and Public Affairs at Tufts University and an aide to the Wisconsin State Legislature.

Brock Evans is vice-president of the National Audubon Society and former head lobbyist for the Washington office of the Sierra Club. He also served as northwest representative of the Sierra Club.

Christy Foote-Smith is executive director of the Massachusetts Association of Conservation Commissions and has been a municipal conservation commissioner in Massachusetts for ten years.

Lois Marie Gibbs is founder of the Love Canal Homeowners Association and is director of Citizens Clearinghouse for Hazardous Wastes in Arlington, Va.

Allen H. Morgan began his career as an environmental leader in 1952 when he started the Sudbury Valley Trustees, one of the first local land trusts in the United States. From 1957 to 1977, he served as executive vice president of the Massachusetts Audubon Society. He is now the New England representative of the Natural Resources Defense Council and consults for other clients in areas of fundraising, membership, and management.

Judy B. Rosener is the assistant dean of the Graduate School of Management, the University of California at Irvine. She served as a commissioner of the California Coastal Commission for eight years.

About the Editor

Stuart Langton is the Lincoln Filene Professor of Citizenship and Public Affairs and the executive director of the Lincoln Filene Center for Citizenship and Public Affairs at Tufts University. He received the Ph.D. in philosophy from Boston University and has been on the faculties of the University of Massachusetts at Lowell, the Whittemore School of Business and Economics at the University of New Hampshire, and Boston University.

Dr. Langton has held many leadership positions including national president of the United Christian Youth Movement; vice-chairman of the Youth Committee of the World Council of Churches; chairman of the National Conference on Citizen Participation; board member of National Council of Churches, U.S. Refugee Committee; Council for the Advancement of Citizenship; and New Hampshire Association for the Elderly. In addition, he has served as a consultant to more than one hundred nonprofit organizations and government agencies.

He has been involved in the environmental movement for twenty years, serving as a director of two wildlife sanctuaries of the Massachusetts Audubon Society; director of the Lowell Environmental Arts and Science Center; vice-president of the Audubon Society of New Hampshire; and founder and initial director of the New England Environmental Leadership Network.